Diplomacy and digital age
Science, technology and global
digital governance

Mario Torres Jarrín (ed.)

Diplomacy and digital age
Science, technology and global digital governance

PETER LANG

Bruxelles - Lausanne - Berlin - Chennai - New York - Oxford

Library of Congress Cataloging-in-Publication
A CIP catalog record for this book has been applied for at the Library of Congress.

Bibliographic information published by the Deutsche Nationalbibliothek
The Deutsche Nationalbibliothek lists this publication in the Deutsche Nationalbibliografie; detailed bibliographic data is available on the Internet at http://dnb.d-nb.de.

This publication is a collaborative effort between university scholars and international experts, developed within the research group "EU Diplomacy Future 4 Digital Age» Thinking an EU Techplomacy Policy for a Global Governance 4.0.

Universidad Pontificia de Salamanca
Institute of European Studies and Human Rights

ISBN 978-3-631-92697-0 (Print)
E-ISBN 978-3-631-92604-8 (E-PDF)
E-ISBN 978-3-631-92966-7 (E-PUB)
DOI.10.3726/b22286

Peter Lang Éditions Scientifiques Internationales - P.I.E., Bruxelles, Belgique

info@peterlang.com - www.peterlang.com

Table of Contents

Chapter 1

Diplomacy in the Digital Age: Introduction, Reflections and a General Overview

MARIO TORRES JARRÍN AND ANTONIO NÚÑEZ Y GARCÍA-SAÚCO

Abstract: The new digital era has profoundly altered every dimension of human life and the whole geopolitical environment in which diplomacy operates. Three issues seem important when dealing with diplomacy in the digital era: first, the characterization of this new era through its most relevant factors, namely, internet and artificial intelligence (AI); second, a description of the main changes new technologies have introduced in traditional diplomacy and, finally, the role of diplomacy in this digital era.

Keywords: Diplomacy, digital age, artificial intelligence, cyberspace, disruptive technologies

1. Introduction

The new digital era has profoundly altered every dimension of human life and the whole geopolitical environment in which diplomacy operates. Defining this new era as digital does not imply any reductionism. It is true that not all new technologies are fully or partially digital, but all of them are deeply intertwined internally and in their common innovative nature. Names and definitions in certain cases are not so important as the generally accepted meaning they convey and sometimes a characterization or description might prove more useful than an inadequate definition. Three issues seem important when dealing with diplomacy in the digital era: first, the characterization of this new era through its most relevant factors, namely, internet and artificial intelligence (AI); second, a description of the main changes new technologies have introduced in traditional diplomacy and, finally, the role of diplomacy in this digital era.

Throughout modern times, scientific rationality molded the so-called Knowledge Society (Drucker, 1969 and Stehr & Ruser, 2017), influencing at the same time, international relations. Conceived as a rational and cohesive system of verifiable knowledge, modern science quickly adopted an empirical approach aimed at general progress. However, the Industrial Revolution prioritized economic progress, providing the resources needed for research and innovation, promoting and expanding scientific development to a level hitherto unknown.

This development had a significant impact on diplomacy, contributing to the emergence of multilateral diplomacy. This book makes historical references to early international organizations associated with telegraphy and radiotelegraphy, reflecting the technical and scientific orientation which the international organizations still preserve (health, climate, biodiversity, energy, nuclear safety, food, water, civil aviation, outer space etc.)

The close interrelation between technical science and economy, fueled by the Industrial Revolution, fostered applied science and specialization. The combination of all these factors has resulted in the technological progress of our time.

Rooted in the realms of data accumulation and instantaneous communication, the internet, social media, artificial intelligence (AI) and automation have propelled the transition from the *Knowledge Society* (Drucker 1989, 1993) to the *Information Society* (Duff, 2000; Brooks, 1994), and from the age of the *Industrial Revolution* to the *Technological Revolution* (Kuehn & Porter, 1981).

Understanding our digital era in its entirety requires an understanding of its global dimension, inseparable from the internet and social media. If internet has facilitated access to information and communication at a global level, social media have amplified both, creating a space of socioeconomic and intercultural exchanges.

Furthermore, the world has been surrounded by new virtual meridians and parallels, delineating all geographical points on our planet, with no place outside their reach. Innumerable access and exit points are permanently available for everyone everywhere, permitting millions of people to express their opinions and feelings on both real and unreal events, and to interact with each other.

Thus, the *Information Society* has generated an immense and constant flow of data whose inherent statistical nature grants them inestimable commercial, political and social value.

Private corporations can take advantage of users' tastes and preferences contained in the data to design productive and commercial strategies, aimed at influencing private consumption to the extent of putting at risk individual privacy.

Public authorities, in addition to or instead of using collected data information to promote better policies, can be faced with authoritarian temptations.

In the social dimension, opinions and emotions expressed in the media, channeled through echo chambers and filter bubbles, can culminate in the construction of powerful narratives that may be used by different groups to exert influence or pressure on national and international decision-makers.

Together with globalization, the second component of the digital era is technologization and automation, i.e, the widespread application of technology in an increasing number of sectors to gain efficiency and the substitution of human work by machines and computerized systems. While globalization is associated with the internet and social media, technologization and automation are linked to AI.

However, all of them are interrelated in a complex way. For their development and functioning, AI and automation require the enormous amounts of data available on the internet. In addition, the vast majority of their applications is done digitally, meaning that both AI and automation depend on the internet and digital technologies to operate.

AI is defined for its technological capacity to artificially perform intelligent functions typically associated to human intelligence. This is achieved through algorithmic automation. Thus, machines, by different methods, have learned to analyze data, recognize regularities and patterns, develop models and simulations, perform sophisticated calculations, anticipate trends, process written and spoken language as well as audio and visual information and contents, make decisions and perform task autonomously.

In addition to these thought-related functions, AI, through automation and robotics, can develop and perform material tasks, both routine and repetitive as well as special and unique ones, including those which require high specialization and extreme precision of execution.

Finally, in its specifically creative function, and beyond its impressive achievements in the cultural and recreative fields, AI has managed to construct a factitious world almost indistinguishable from the real one, a capability akin to the internet.

Internet has woven a virtual universe around narratives, mentally superimposed on and often eclipsing the real one. Generative AI has fabricated spaces where the lines between the real and the unreal are obliterated. While the internet has created an enveloping and encompassing experience, AI provides an immersive and intrusive one. But both potentially threaten the clarity of truth and the perception of reality.

Furthermore, global digitalization has introduced new barriers while eliminating others, all of which has significantly impacted diplomacy. Varying levels of access to the internet and social media have created additional social divisions while amplifying previous ones in and outside states. The United Nations has identified numerous groups whose access to the internet is limited, reducing social, political, and economic opportunities.

Given the deep technological gaps between states, automation and robotics have also contributed to reinforcing international inequality. Forecasts indicate its continuous increase due to the extraordinary speed at which technological innovation advances.

Consequently, the digital era has shaped a new international environment characterized by growing and reinforced imbalances and a hitherto unknown selective concentration of global influence. However, while new borders have been erected, others have been removed and dismantled.

As previously mentioned, spatial barriers have been blurred both materially and immaterially. Materially, the necessity for physical co-presence or face-to-face diplomacy has disappeared. From an immaterial perspective, universal and local, international and national dimensions have blended, affecting the concept of territorial sovereignty, the cornerstone of state power and national identity.

Moreover, the digital age has eradicated temporal limits. Time has traditionally played a pivotal role in diplomatic practices. Thoughtful and strategic time management, prudent decision-making and wise planning were invariably associated with classical diplomatic skills. However, in the global digital age, diplomacy is increasingly pressured towards immediate responses and reactions, escalating the risks of errors with far-reaching global consequences.

Finally, as said before, the digital age has transcended the confines of the real world, establishing an alternate realm beyond tangible reality—a new virtual universe brimming with immense innovative potential.

Navigating and operating in this changing, fragmented, dual-sided, hybrid, and contradictory environment represents one of the paramount challenges diplomacies must face in the digital era.

2. Reflections

Our current digital age has introduced substantial advantages and benefits, alongside significant risks and serious dangers. Traditional diplomacy can be understood by considering key actors and their functions as well as its conventional agenda and operational methods. States, limited in number and sovereign within well-defined geographical boundaries, were the exclusive subjects of international law and, alongside intergovernmental organizations, the primary diplomatic actors.

With well-defined functions, sovereign and independent, states agreed to engage formally with one another only within the framework of strict traditional or agreed-upon rules and methods. In short, traditional diplomacy has operated within a subtle yet meticulously protocol-regulated space, where one of the few discretionary aspects involved time management skills.

In this historical framework, the fundamental functions of traditional state diplomacy were threefold: active and passive diplomatic accreditation; negotiation with other states concerning shared or general interests (bilateral or multilateral diplomacy); and, thirdly, the defense and promotion of national interests. The conventional diplomatic agenda was focused on national interests, routinely linked, but for exceptional cases, with the handling of issues arising from the different domestic institutions, on which states heavily relied to sustain their diplomatic activities. However, this traditional paradigm has undergone a dramatic transformation in the digital era.

Firstly, the factual state monopoly in international relations has diminished as globalization has exponentially increased the number and significance of new international actors, both inside and outside the state framework. Within the state, not only state powers and governmental departments, but also sub-state entities at regional or municipal levels are increasingly engaged in their own international cooperation programs.

Beyond the state's borders, intergovernmental organizations have gained expanding importance alongside globalization, frequently constraining or supplanting the diplomatic actions of individual states. This book particularly emphasizes the leadership assumed by the European Union (EU) as a supranational organization, a function individual states would not have been able to undertake independently.

Outside the scope of state influence, civil society undeniably exerts a leadership role on the international stage. Private associations, institutions, foundations as well as scientific and technological entities of renown have gained significant force. As an example, the *Swedish Royal Institute of Technology* (KTH) is showcased in the publication at hand.

Moreover, various groups and even individuals have acquired substantial global influence. Numerous well-known groups possess the capability to exert pressure on both national and international levels, while certain individuals leverage their prestige to promote supranational initiatives. The case of Dr. Adi Anani, as mentioned in a chapter of this book, exemplifies the so called "individual diplomacy" by facilitating relations between Swedish and Palestinian scientific institutions.

However, major digital and technology companies such as Alphabet, Amazon, Apple, Facebook, Google, Meta, Twitter (X), Microsoft in US, or Alibaba, Baidu, Tencent or Xiaomi in China stand out above all other new international actors as they have contributed more significantly than any other entity to shaping the geopolitical environment in our age. These Big Tech Companies are emerging as geopolitical actors in their own right (Torres Jarrín & Riordan, 2023). Their importance is reflected in the recurrent references made to their role throughout this book.

In conclusion, states and governments do not exclusively cover the global scope of international relations. The traditional notion of diplomacy as the primary and sole responsibility of states has given way to the involvement of numerous sub-state and non-state actors capable of wielding substantial influence at an international and global level.

Similar observations can be made of the fundamental state functions, the traditional diplomatic agenda as well as its working methods.

Accreditation, once an exclusive symbol and seal of a state's sovereign identity and an indispensable formality for recognition as legitimate international law subject, has been diluted. Ambassadors have been diplomatically accredited to various major technology companies in Silicon Valley recognizing them on equal footing with states. Mark Zuckerberg has reflected this unequivocally when he said that in many ways Facebook is more like a government than a traditional company.

The second classic function of the state involved the exclusive authority to enter into legally binding agreements with other states as equals, along with the privilege of selective participation in international diplomatic conferences. Both functions have undergone substantial changes. The expansion of international agreements between states and non-state actors has continued. Similarly, the involvement of civil society organizations in international diplomatic conferences, such as the Paris UN Conference on Climate Change, has seen significant growth.

The third traditional state function of defending and promoting national interests has gradually been overshadowed by the aforementioned private entities with greater capacity to advocate for their own interests.

Concurrently with this process of privatization, globalization has eroded every clear distinction between national and international issues. Nothing can remain beyond the scope of globalization which has permeated every aspect of modern life with a supranational dimension. Consequently, states are no longer the exclusive and complete masters of their diplomatic agendas.

Finally, the instantaneous and universally accessible nature of information has disrupted the traditional confidentiality and discretion that formerly safeguarded classical diplomacy. This shift has subjected diplomacy to the hazards of excessive exposure to media and public opinion.

Today, no state can ensure diplomatic secrecy: quantum computers could be able to decrypt any encryption. How can diplomacy adapt to these changes?

The least risky approach would be to briefly consider the closest analogous precedent, namely how science and diplomacy converged into *Science Diplomacy* (Flink & Schreiterer, 2010; Riordan & Torres Jarrín, 2020). This seems adequate given the significant role of science in today's international landscape and the inherent relationship between science and technology.

The London Royal Society and the American Association for the Advancement of Science are known to have condensed the mutual interconnection between science and diplomacy into a few maxims, such as science *in* and *for* diplomacy and vice versa. A similar approach can be adopted in the case of modern new technologies, posing the same initial question: should one take preference over the other?

The answer to this question depends on the utilitarian or final character that can be conferred to one over the other. Science and technology inherently tend to evolve and expand. In fact, both have constituted the foremost universal and global phenomena in history, finding a fertile ground within the international nature of diplomacy. Consequently, diplomacy has been and continues to be an integral factor in their internal development and a channel for their external dissemination.

However, it is evident that diplomacy is not merely a conduit or tool in areas where its central activity is fundamentally scientific or technological. For example, negotiating the regulations governing internet usage is an active task that unmistakably extends beyond any simplistic interpretation of diplomacy as a mere tool (*diplomacy in science/technology*).

Likewise, solving conflicts, fostering cooperation or securing beneficial agreements in the realm of science and technology represents a natural function of diplomacy (*diplomacy for science/technology*). In those instances, diplomacy assumes the role of an actor.

There is also the reverse question of whether science/technology has an instrumental or a final significance in state diplomacy. States and governments can undoubtedly leverage science and technology much like they utilize culture or art to advance their national interests on the international stage or turn to the assistance of science and technology to support decisions on diplomatic matters (*science/technology for diplomacy*).

3. Conclusion: A General Overview

But let us come back to the main question: what is the primary role of diplomacy in our digital era? There is no doubt that regulating new technologies, their use and their impact has become its central function and ultimate objective. However, where should we begin?

The enormous complexity of the task requires a clear definition of criteria. As UN experts have warned, given the diverse global contexts and risk landscapes, sustainable solutions demand a customized approach. Nevertheless, for our discussion at a general level, we should focus on which technological aspect or sector should be prioritized for regulation.

At the dawn of the digital era, the question of how to utilize the internet and integrate it into diplomatic practices was perceived as important (*eDiplomacy*).

With the expansion of digitalization and its media use, the need for better defined and more consistent relationships between diplomacy and digital networks was considered a priority (*Digital Diplomacy*).

When the first signs of the previously little-feared risks and dangers linked to modern technologies began to appear, diplomacy sought the cooperation of the major technology companies to try to avert them (*Techplomacy)* (Torres Jarrín, 2022; Torres Jarrín & Riordan, 2019).

Now, with the emergence of generative intelligence, it seems that the priority lies in regulating AI, while maintaining a constant and special attention to cyberspace (*Cyber Diplomacy*) (Riordan, 2019), as it encompasses all online activities, the entirety of the internet, and virtually all AI operations. Moreover, if regulation needs to prioritize a risk-based approach, cyberspace remains a major and persistent threat.

Well-documented dangers of AI are i.a., cyberattacks, interferences in electoral processes, misinformation, social discrimination, political radicalization and polarization, direct violations of personal privacy (*intimacy intrusion*). Today we live equally in a society of information and disinformation (Nuñez y García-Saúco, 2020). AI can also indirectly affect privacy through computer-human interaction (CHI) and by using subtle methods to tempt users to reveal personal data, accept companies′ preferred options, retain loyalty as long as possible etc.

The potential dangers, on the other hand, include a broad spectrum, ranging from lethal autonomous weapons capable of independent operation without human involvement, to a direct threat to human intelligence, well-being, and even existence.

World-renowned scientists, including Stephen Hawking or Geoffrey Hinton have expressed serious concerns about the potential risks of recent most advanced technologies. Other experts, like Sam Altman, director of Open AI and designer of ChatGPT, have acknowledged fears of large-scale disinformation and cyberattacks. Mo Gawdat, former Chief Business Officer of Google, has described AI's potential dangers as beyond any previous emergency.

In 2023, around a thousand signatures, including some of the world's most prestigious scientists, called on all AI laboratories to publicly and verifiably halt the training of all systems more powerful than GTP-4 for at least six months. This pause would allow for reflection until society could regain confidence that AI's effects will be positive and its risks manageable.

In view of such numerous and noteworthy testimonies, it seems evident that, following the growing sense of uncertainty and insecurity created by the most recent technologies, the immediate diplomatic priority has shifted towards regulating damage prevention and risk management.

The need to promote the necessary regulatory processes turns out to be the preeminent purpose of contemporary diplomacy, aiming to achieve two main objectives. The first and most immediate one is to prevent and mitigate real dangers and potential threats inherent in new technologies. Secondly, in the medium term, the goal is to instill transparency, responsibility, reliability, security and protection permitting the maximization of their benefits. However, in parallel to the manifold efforts, attempts and projects, both at the level of states and inter-state organizations as well as private and international entities, regulation has been subjected to a double contradictory pressure. On the one hand, there was a widespread call for swift regulation. On the other hand, as mentioned, numerous appeals demanded to slow down, even stop, not only regulation, but also technological research developments.

This concern about premature regulation leads to the argument that new technologies, still evolving, have not consolidated sufficient maturity to allow for a scientific assessment of the true nature and scope of their many effects and consequences. Regulating AI at this stage of evolution could eventually bring more harms than benefits.

Furthermore, strict regulation has scientific and economic implications that could seriously slow down innovation itself. There is also growing awareness that achieving fully effective regulation will never be possible because of the natural pace of scientific and technologic innovative advances, compared to the usual slowness of political and diplomatic decision-making process. Regulation will inevitably struggle to keep in pace with the latest scientific and technological advancements, thus remaining partial and outdated. This dilemma is reminiscent of the paradox between Achilles and the tortoise.

The analogy is further emphasized by the assessment of diplomacy as not yet fully equipped to successfully address new digital challenges.

All of the above may have induced a part of public opinion, especially in the US, to appeal to the responsibility of companies to regulate themselves. This appeal goes hand in hand with the idea that any attempt at public regulation could potentially infringe upon the First Amendment which guarantees freedom of speech and freedom of the press. The appeal is in fact a call for ethical and social corporate responsibility, exemplified in leading companies such as IBM or Microsoft, among others, that have equipped themselves with corporate ethics guidelines.

Companies can significantly contribute to avoid and mitigate a large number of dangers like data breaches or privacy violations by integrating such guides into their company culture. However, who can guarantee that in a conflict of interest's ethical values will prevail over economic benefits when the company itself is judge and party. Regulation cannot be completely left in the hands of companies. The dilemma of greater or lesser urgency of regulation cannot bypass addressing some basic questions. As UNESCO has indicated, finding the best way to regulate is as important as regulation itself (UNESCO, 2022). Three criteria indispensables for efficient regulation are the method of implementation, key stakeholders' involvement and the underlying framework.

Regarding the first criterion, there is general consensus that regulation must adopt a synergistic approach integrating all aspects are to be covered (technical, legal, political, social, economic and ethical), balancing them with the need for innovative progress.

The technical aspects are mainly related to algorithms. In theory, they offer the advantage of operating objectively, without being affected by human emotions. They can rapidly analyze vast amounts of data as well as identify and flag possible risks. In addition, they can be designed to adapt to the evolving nature of technologies, automatically adjusting to new developments. As a consequence, algorithms themselves are sometimes being regarded as regulatory or co-regulatory instrumental agents. However, the grave potential consequences of modern technologies cannot be left to blind, opaque and unaccountable algorithms. While their capacities can be applied in the regulatory process, the responsibility must remain in human hands.

The entire regulation is inherently a legal process. Nevertheless, there are specific legal aspects which deserve special attention, such as algorithm biases, privacy rights, personal data protection, right to be forgotten, freedom of speech, transparency obligations, fair competition, consumer protection, intellectual rights, taxation, cybersecurity and liability and accountability.

The political dimension is linked to the defense of basic democratic values. Regulation cannot solely target illegal election interference; it must encompass all necessary aspects to prevent the extraordinarily powerful technology companies, given their lack of representative legitimacy, from exerting undue influence on public opinion.

The regulation of social aspects overlaps, on the one hand, with similar political phenomena, such as misinformation, expressions of hate, inequality, fragmentation, polarization, confrontation. On the other hand, it requires attention to technical aspects, such as ensuring that datasets used for training algorithms are socially representative of the diversity of the target population to avoid structural biases in algorithms. Additionally, the widespread current concern about the medium-term impact of automation and robotics on labor markets needs to be addressed.

The privileged and dominant economic position of big technology companies, along with the profound changes in production and trade systems, call for new regulations in several key economic domains. These include, i.e., fair competition, user and consumer protection, e-commerce, digital banking and currencies and equitable digital services taxation.

Finally, the ethical aspect presents a general and a practical dimension. Ethical values must, as a principle, permeate all regulation aspects. In its practical dimension, the OECD AI Principles through of the (Recommendation of the Council on Artificial Intelligence, 2019) or the Guiding Principles on AI, and more specifically, the voluntary Code of Conduct for AI Developers, recently approved by G7 leaders, serve as exemplary instances. The Code of Conduct is based on a few important specific requirements that developers adhere to voluntarily, aiming to guarantee transparency and responsibility throughout the life cycle of each AI system (Hiroshima Process International Code of Conduct for Organizations Developing Advanced AI Systems, 2023). From an European perspective, the EU´s approach to AI centers on excellent and trust, aiming to boost Research and industrial capacity while ensuring safety and fundamental Rights (European Commision, 2018). The use of AI in the EU is regulated by tue EU AI Act (European Parliament, 2024), that is the world´s first comprehensive AI law. This act is part of the EU digital strategy, the EU wants to regulate AI to ensure better conditions for the development and use of this disruptive technology.

The second criterion for adequate regulation, in parallel with the synergistic requirement, emphasizes also an inclusive approach, requiring the involvement of all main actors. Building on this concept, a comprehensive and cooperative model should start with a bottom-up approach, prioritizing input from scientists and experts who work directly with algorithms, models and systems. This ensures the initial adequate effectiveness by leveraging technical expertise. Additionally, respected and influential experts and scientific organizations, public or private, dealing with the different spheres mentioned above (e.g. ethics, social, legal) should be included.

The participation of leading technology companies, as well as industry representatives, is indispensable. There is, as has been said, no shortage of companies that have already adopted ethical codes of conduct or regulation protocols that include practices of social corporate responsibility. Incorporating useful and successful experiences of the industry sector can both facilitate and enrich the regulation process.

As the primary subject of the benefits and perils of new technologies and as the ultimate recipient of the final regulation results, civil society stands as a crucial stakeholder. Its participation embodies public concerns as well as social and ethical priorities and ensures that these considerations are at the forefront of the regulation process.

However, an efficient regulatory model requires a complementary top-down approach. In this regard, state participation becomes a decisive factor. Non-state actors can wield an undeniable soft- power influence. But this influence does not confer diplomatic status upon them, unless integrated into state diplomacy. Only states possess the power and legal authority to coercively legislate and create binding regulations.

With this in mind, and considering the need for cooperation, intergovernmental organizations has emerged as relevant and efficient instances bringing together stakeholders and promoting regulation. The African Union, ASEAN, G-7 Group, the Council of Europe and, above all, the European Union have already taken important decisions in this regard. However, the global nature of the issues to be regulated and the need to convene all the main actors, requires a framework that transcends both regional and specialized intergovernmental organizations. Only the UN has the inclusive and legitimate power to coordinate the needed global approach. intergovernmental organizations can help in many ways, but the final role of the UN is irreplaceable.

In view of the enormous challenges facing diplomacy in the digital age, the ultimate question is: will diplomacy succeed in achieving these ambitious goals? Despite being described as multipolar, the new international order remains largely bipolar due to the confrontation between the United States of America (USA) and the People's Republic of China, evidenced by two key factors: firstly, USA and China stand as the two primary and most influential actors in the current global-digital order; second, the majority of this diplomatic confrontation occurs, either directly or indirectly, within the realm of science and technology, particularly in the arena of the new high technologies.

The USA, despite its recently increased regulatory efforts, still lacks a unified and coherent regulatory framework. China has a comparatively more consistent and comprehensive, periodically updated, but still limited system. The USA follows a clearly liberal socioeconomic model, where business profit takes precedence over social interest. In China, domestic policy enforces tight control over public opinion, with this book providing some illustrative examples. However, no global agreement is possible without the participation of the USA and China.

In contrast, the European Union (EU) is founded on the principles of the social rule of law, where both individual rights and collective welfare hold equal importance. Based on these values, the EU has assumed a global influential role in the regulation of new technologies. With the USA, China and the EU as the main leading global actors, the future of diplomacy will depend, firstly, on how the USA-China confrontation evolves and, secondly, how the EU will manage to maintain its leadership.

Predicting the path Which the Sino-American rivalry will follow seems an extremely difficult task. The inherent contradictions within and between USA-China's respective models in social, political and economic terms, the differences in the regulatory approach, coupled with the intense competition to impose each one's own technological supremacy, obstruct the much-needed basic cooperation between them for the foreseeable future. At the same time, as well noted in the book, the EU has to take important internal and external decisions to achieve its goals.

Internally, the EU lacks the major technological companies that USA and China possess. Consequently, the EU must make an enormous financial effort to support research in highly competitive sectors.

Externally, the EU needs to attract other states to its cause, particularly those with which it shares its democratic values, starting with the USA. A strong alignment between the two would reinforce each other's assets and bolster the position of the West vis-a-vis China. Other countries such as Australia, Canada, Japan, New Zealand and South Korea represent an added value, as does Latin American, a continent on which this book has placed particular emphasis, and to whose democratic development the EU has been specially committed for many years through technical technological cooperation.

It is surprising that the EU, a global actor without big technology companies, has been the major force spearheading regulation. Perhaps this is precisely due to greater freedom from undue influence.

References

Brooks, H. (1994). The relationship between science and technology. *Research Policy,* 23(1994), 477–486. Elsecier Science B.V. https://www.belfercenter.org/sites/default/files/legacy/files/sciencetechnology.pdf

Drucker, P. F. (1969). *The age of discontinuity. Gidelines to our changing society.* New York: Harper & Row.

Drucker, P. F. (1986). *The new realities.* London and New York: Routledge.

Drucker, P. F. (1993). *Post-capitalist Society.* Oxford: Butterworth-Heinemann Ltd.

Duff, A. (2000). *Information society studies.* London and New York: Routledge.

European Commission. (2018). Communication from the Commission to European Parliament, the European Council, the Council, the Eruropean Economic and Social Committee and the Committee of the Regions. Artificial Intelligence for Europe. COM/2018/237 final. https://eur-lex.europa.eu/legal-content/EN/TXT/?uri=COM:2018:237:FIN

European Parliament. (2024). European Parliament legislative resolution of 13 March 2024 on the Proposal for a regulation of the European Parliament and the Council on laying down harmonised rules on (Artificial Intelligence Act) and amending certain Union Legislative Acts (COM (2021) 0206—C9-0146/2021-2021/0106(COD)) https://www.europarl.europa.eu/doceo/document/TA-9-2024-0138_EN.html#title2

Flink. T. & Schreiterer. U. (2010). Science diplomacy at the intersection of S&T policies and foreign affairs: Toward a typology of national approaches. *Science and Public Affairs*, 37(9), November, 665–677.

Hiroshima Process International Code of Conduct for Organizations Developing Advanced AI Systems. (2023). European Commission. https://digital-strategy.ec.europa.eu/en/library/hiroshima-process-international-code-conduct-advanced-ai-systems

Kuehn, T. & Porter, A. L. (1981). *Sience, technology, and national policy.* Ithaca, New York: Cornel University Press.

Núnez y García-Saúco, A. (2020). Old and new order. In G. L. Gardini, *The world before and after COVID-19. Intellectual reflections on politics, Diplomacy and international relations* (p. 31). Salamanca-Stockholm: European Institute of International Studies Press.

Recommendation of the Council on Artificial Intelligence. (2019). OECD Legal Instruments. https://legalinstruments.oecd.org/en/instruments/OECD-LEGAL-0449#supportDocuments

Riordan, S. (2019). *Cyberdiplomacy. Managing security and governance online.* Cambridge: Polity Press.

Riordan, S. & Torres Jarrín, M. (2020). Science diplomacy. In *Global policy perespective report.* Salamanca-Stockholm: European Institute of International Studies Press.

Stehr, N. & Ruser, A. (2017). Knowledge society, knowledge economy and knowledge democracy. In E. Carayannis, D. Campbell, & M. Efthymiopoulos (eds.), *Handbook of cyber-development, cyber-democracy, and cyber-defense.* Cham: Springer.

Torres Jarrín, M. (2022). Techplomacy. El interregionalismo de la Unión Europea-América Latina y el Caribe en la era digital: Un interregionalismo 4.0 para una Gobernanza 4.0. In B. Larraín & P. Astroza (eds.), *Relaciones entre la Unión Europea y América Latina: Escenarios futuros en un mundo de cambios.* Valencia: Tirant Lo Blanch.

Torres Jarrín, M. & Riordan, S. (2019). *Techplomacy. Hacia la búsqueda de una regulación del ciberespacio y la gobernanza de Internet, en Beltrame de Moura. O Direito Internacional Privado Europeu: entre a harmonização e a fragmentação, Emais Editora y Librería Jurídica.* Brasil: Publicación financiada el Programa Erasmus + de la Unión Europea.

Torres Jarrín, M. & Riordan, S. (2023). *Science diplomacy, cyberdiplomacy and techplomacy in EU-LAC relations. United Nations university series on regionalism*, vol. 25. Switserland: Springer.

UNESCO. (2022). *Recommendation on the ethics of Artificial Intelligence.* Paris: United Nations Educational, Scientific and Cultural Organization. https://unesdoc.unesco.org/ark:/48223/pf0000381137

Chapter 2

AI and International Relations: Building a Practical Guide for Ethics Governance of AI for Diplomacy

Cecilia Danesi

Abstract: The use of AI has revolutionised many sectors in our societies: juridical, economic, social and cultural. In international relations terms, it represents a big challenge for governments because it includes a radical change of mind-set. Diplomacy is being part of this transformation from two main different sides: AI as a tool for diplomacy and AI as a topic for diplomacy. This obligates diplomats to be aware of AI and particularly, to lead their job with an ethics governance of the AI perspective. This paper addresses the main critical areas that diplomats have to take into special consideration such as transparency/explainability, algorithmic bias, AI regulation (with focus on AI Act), autonomous weapons, Gen-AI and disinformation and democracy. Finally, the chapter gives the key points of the "Practical guide of AI Ethics principles in Diplomacy".

Keywords: Artificial intelligence. Diplomacy. Foreign affairs. Human Rights. Ethics. Governance. Algorithmic bias. AI Act. International relations. Autonomous weapons. Disinformation. Democracy. Generative AI

1. Introduction

We can easily affirm that artificial intelligence, the "main character" of the Fourth Industrial Revolution, is high almost everywhere; in fact, it has become part of our life. We can also agree that artificial intelligence has several advantages for our society. For example, to prevent natural disasters, to diagnose diseases, to improve customer services or to manage social media.

Diplomacy does not escape this ubiquitous AI to the emerging technologies. The big proof is the decision of several countries to create "Tech ambassadors", such as the case of Denmark, pioneer in this field. The main objective of them is to focus on the relationship between governments and huge tech companies, for that reason, they are based on the main stages of where tech is being developed: Silicon Valley, Beijing and sometimes in East Africa. This is a clear way to recognize not only the power of tech companies as international actors but also, recognize its power as big as States, with a kind of sovereignty.

As we know, the emerging technologies never stop. Recently, we've begun to face another new challenge: the Metaverse, where all these technologies are found and growing together. This new context has just arrived to the diplomacy area: "Barbados is opening a diplomatic embassy in the Metaverse" was stated in the news.

What are the risks and advantages of these emerging technologies for diplomacy? How do they influence in Human rights and ethical concerns? Do we need a legal framework? Let's see.

2. What Is Artificial Intelligence?

AI is not a new phenomenon. In fact, this word appeared for the first time in 1956 when John McCarthy, an informatic scientist, used it in the famous Dartmouth Conference. But the idea that machines can emulate our brain was older; Turing (1950) and Claude Shannon (1938), for example, made big contributions in the field before this conference. The reason why now artificial intelligence has become a buzzword is thanks to the big amount of data that we produce every day and the ability and capacity of processing it by computers.

Defining AI is not easy. There are several disciplines involved working on it. In a nutshell, autonomous systems make predictions taking into account data from the past (a group of data which is called a "data set"). To do that, the data is processed by different algorithms, which are like cooking recipes; that is to say, a sequence of logical steps.

The big revolution inside AI is machine learning, which means (as we can read) a computer that can learn by itself. From the 1950s to the 1980s, the paradigm of symbolic artificial intelligence prevailed. Programming was called classical under the "*If…then…else*" rule. This consisted of humans having to enter rules and data into a program and, from the analysis of the data under those rules, the expected answers would result (Moreno et al., 1994). But machine learning does not need rules, it just needs data, and will create the rules taking said data into account.

Machine learning, in a broad definition, includes any computer program that improves its performance at some tasks through experience. In a more technical way, "A computer program is said to learn from experience E with respect to some class of tasks T and performance measure P, if its performance at tasks in T, as measured by P, improves with experience E" (Mitchell, 1997).

Nowadays, the study of the challenge, risk and advantages of AI is in the international agenda of Governments, companies and associations, for that reason, we can find different documents which provide a concept of that. The High-Level Expert Group on Artificial Intelligence expressed that artificial intelligence are systems that "given a complex goal, act in the physical or digital dimension by perceiving their environment through data acquisition, interpreting the collected structured or unstructured data, reasoning on the knowledge, or processing the information, derived from this data and deciding the best action(s) to take to achieve the given goal. They can also adapt their behaviour by analyzing how the environment is affected by their previous actions."

Later, the European Parliament issued a proposal for a regulation laying down harmonised rules on artificial intelligence called "AI Act" which also provided a definition. Taking into account the latest draft agreement "An AI System is a machine-based system designed to operate with varying levels of autonomy and that may exhibit adaptiveness after deployment and that, for explicit or implicit objectives, infers, from the input it receives, how to generate outputs such as predictions, content, recommendations, or decisions that can influence physical or virtual environments".

As we can see, the two concepts are different but the aim is similar, i.e. both highlight the main characters of AI systems. The first one is more detailed because it specifies its technical operations. In any case, that main characters are the definition of an objective by a human (for example, detect lung cancer in a chest x-ray), the capacity of processing large amounts of data[1] and the high speed in doing so, self-learning capabilities and its controversial "autonomy", which means that AI can act (make a decision) without any external control.

---

1 For that reason, nowadays we talk about the "data economy" and the big power of data. To see the consequences of the loose of privacy due to the "Data economy", Véliz, Carissa, Privacy is Power: Why and How You Should Take Back Control of Your Data, Bantam Press, 2021.

To conclude this paragraph, we can easily affirm that nowadays power relies on data. For that reason, we talk about "data economy" and some well-known authors mention the rise of "Data Religion" which states that the entire universe is a flow of data and that organisms are algorithms; all of this due to the data-processing technologies (Harari, 2016).

3. The Relationship Between AI and Diplomacy

The impact of artificial intelligence in diplomacy is huge and deep; there are several areas involved. The reason is simple: AI is part of our lives, a part of citizenship's life, and is a great tool to improve processes and tasks.

Technology is already inside diplomacy. "Science diplomacy" is defined as "the use of scientific collaborations among nations to address the common problems facing 21(st) century humanity and to build constructive international partnerships" (Fedoroff, 2009). Some authors consider that science diplomacy links the two policy domains of foreign affairs and science policy (Leijten, 2017) and others state that it has to be included in that concept "science, technology and innovation diplomacy" (ST&I) (Araújo-Moreira et al., 2022).

In addition, the idea of "cyber diplomacy", originated by the irreversible process of digitization is linked with our field of study. This area consists of "the use of diplomatic tools to address issues originated in and through cyberspace. Those issues cover a plethora of different aspects like security, economic and human rights topics, including international cybersecurity standards, internet access, privacy, internet freedom, intellectual property, cybercrime, ethical use of digital technologies, among many others" (Araújo-Moreira et al., 2022).

Focusing on AI, we can divide its impact into Diplomacy in three big groups: AI as a topic for diplomacy, AI-related shifts in the environment in which diplomacy is practiced and AI as a tool for diplomacy (Höne et al., 2019). We believe that it is better to study the last two groups together due to their close links between them.

3.1. AI as a Topic for Diplomacy

Firstly, AI as a topic for diplomacy refers to the importance of AI in geopolitics, governance, economy, just to mention a few. Proof of this is the number of recent documents published by States, cities, international organizations addressing topics related to emerging technologies and, specially, autonomous systems. Therefore, diplomats should have a complete knowledge about the phenomenon to carry out their role.

One of the most important tasks that diplomats must carry out is the promotion of common global regulatory frameworks. As we know, AI does not admit borders; no emerging technology does. Let's take for example the Uber case. This company changed the mobility of cities, without asking for permission and without -at first- respecting the law. As if all this was not enough, Uber faced one of the most powerful unions: taxi employees. Unthinkable a decade ago, right? For example, due to the Uber impact and the pandemic, the request for taxi license renewal this year was less than a half that of previous years in Buenos Aires city, Argentina.

Another point of analysis is the role and power of big tech companies in the international scenario. The actions of technology companies linked to AI have a geopolitical impact worldwide. Their economic capacity and their power to influence societies make them, de facto, new actors within the international system and global governance (Torres Jarrín 2024). As we mentioned in the introduction, one clear example is the birth of Tech Embassies.

We are going to study deeply some special aspects of this topic in Section 4.2.

3.2. AI as a Tool for Diplomacy

AI as a tool for diplomacy is linked to the use of it as an assistant of diplomats' work. International stakeholders with advanced AI can exercise more political, military, economic and diplomatic power. Governments around the globe are seeking to harness AI's potential for economic advantage, technological supremacy and influence over global norms and standards, but specially, the proliferation of GenAI raises a myriad of economic, political and ethical issues for diplomatic relationship-building (Qian Xia, 2023).

There are several examples of the use of AI in diplomacy. Firstly, through chatbots (which are systems that use natural language processing to have conversations with humans, seeming to be one of them).

Frequently, a lot of banks or companies use chatbots as a form of customer service to resolve issues. This could be used for consular tasks.,

Secondly, AI could be used for negotiations and designing action strategies. This operation is not new for autonomous systems. There are several cases where it serves as a tool to, for example, interpret and summarize contracts and also to propose strategies for negotiation. That is the case of Speed Legal, a software that is used for analyzing contracts by lawyers and consumers. It automatically identifies red flags that do not comply with the market standard and establishes the risks of the terms and finalizes the reports to present the data to the top management. Also, the AI advantages in assisting in negotiations have been proved with football player Kevin De Bruyne, who signed an excellent renewal of his contract with Manchester City thanks to Big Data's advice without any human agent. Then, another football player, Esteban Granero created a football-based Artificial Intelligence consultancy firm called Olocip, which can predict how good a player will be; information that can be used to negotiate a contract.

By processing past data, AI has the ability to predict the future with (in many cases) great accuracy[2]. Thus, artificial intelligence is a useful tool for carrying out diplomats' tasks. Algorithm design and data mining can help improve understanding and management of complex scenarios and a lot of tasks: prediction of migratory flows, geo-coding and mapping in social networks to measure the impact and effectiveness of political messages, the management of migratory crises in sensitive borders, among others[3].

Nevertheless, apart from these indisputable advantages, AI has several dark sides. Especially when we talk about humans' rights. Let's see.

---

2 In its technical meaning, the "accuracy" of a model is a percentage that let you know how exact the predictions are. To train it, you will separate the dataset in "train data" and "test data". With the first one, the system will learn. Then, you will evaluate how it learns with the "test data" and, during this process, the system will give you a percentage of accuracy, which indicates the margin of error.

3 See more in "Las transformaciones de la diplomacia", N.º 6 Oficina de Estrategia y Prospectiva, Ministerio de Relaciones Exteriores del Gobierno de España, December 2021, available in https://www.exteriores.gob.es/es/ServiciosAlCiudadano/PublicacionesOficiales/En%20prospectiva,%20n%C2%BA6.pdf (Accessed 10 October 2022).

4. Why Should Diplomacy Be Awareness of AI? Critical Areas

A clear summary of AI risks can be found in almost all Ethical AI principles, especially in the Ethics Guidelines for Trustworthy AI of the European Union and the Recommendation on the Ethics of Artificial Intelligence by UNESCO. Considering the diplomat's role, we would like to highlight the main critical areas: transparency/explainability, algorithmic bias, disinformation, and the use of AI in war. Let's explore each of them.

4.1. Transparency/Explainability and Algorithmic Bias

Two issues that are particularly troubling. The first is AI transparency or explainability, and the second is algorithmic bias. The former refers to the inability to explain how a system reaches a decision. This inability can manifest in two different ways: legal and technological. In the first case, the operation of the system cannot be disclosed due to legal barriers, such as intellectual property law or trade secrets. In the case of technological barriers, the lack of explainability is attributed to the model's operation, meaning that even the most expert workers in AI cannot discern which internal processes led to a specific decision. This is typically observed in deep learning models.

To address the lack of transparency in digital platforms, the European Union has issued the Digital Services Act package, which includes the Digital Markets Act and the Digital Services Act (DSA). The latter enforces obligations on online intermediaries and platforms such as online marketplaces, social networks, content-sharing platforms, app stores, and online travel and accommodation platforms. It's worth mentioning that in most cases, these platforms use algorithms to manage content. As a result, the DSA stipulates transparency obligations.

In turn, algorithmic bias appears as a mistake in the system, which leads to a wrongful prediction. This in particular grabs attention in the field of Social Science when the wrongful prediction implies discrimination or an unfavorable treatment.

Both cases have to be the concern of diplomats in the two uses that we mentioned above, AI as a tool for diplomacy and AI as a topic for diplomacy. Algorithmic bias is potentially a very dangerous problem because it is more difficult to identify and to prevent. The way to recognise it is through algorithmic audits and impact assessments that have to be mandatory under law. This is the case of New York City, which enacted a law that requires an annual algorithmic audit of AI systems used in job areas (to screen candidates, for promotion/recruitment, etc.).

Prevention is difficult. Bias is transferred into technology by humans. This could happen through datasets, or even in each decision that people make during the design, development and maintenance of a system. So, it is complex to realise how and where biases appear (Danesi, 2022).

The capacity to control the learning process of a system depends on the kind of model that we use. GPT (Generative Pretrained Transformer), from OpenAI, is a clear example of that. The system is used for writing: you can start a sentence and GPT can complete it; it can also write an article. The problem is that the system learns freely from the internet (Wikipedia, websites, etc)—it can learn both good and bad things and exerting control over it suggests significant challenges.

In the area of diplomacy, the system can be used to create scenarios for future developments; "It would be conceivable, for example, to 'eed' the system with specific information in addition to the general stock of knowledge, such as newspaper reports on the topic or drafts of negotiation texts. By combining the general stock of knowledge with this specific information, the system could then automatically develop a series of scenarios at the push of a button" (Stanzel et al., 2022). This significant advantage for diplomacy cannot be realised without addressing algorithmic bias. Scenario building must not be influenced by biases and stereotypes.

For this reason, data quality and conducting big data analyses raise questions about access, interpretation, protection, and security. These issues remain a key concern in the context of AI and diplomacy, particularly in cases concerning sensitive data or political decisions with potentially far-reaching consequences (Burkadze, 2021).

4.2. AI Regulation: AI Act and Autonomous Weapons

As we mentioned above, diplomacy has a lot to contribute to building a regulation of artificial intelligence that is harmonious and respects ethical principles. First, because we need an AI regulation that is as harmonious as possible to be able to effectively limit the power of technological giants. If countries create isolated regulations, enforcement becomes challenging. . Additionally, they risk losing innovative opportunities and technological advancements. Secondly, while acknowledging the variability of ethics among individuals, significant agreements have been reached concerning AI, exemplified by the UNESCO AI Ethics Recommendation signed by over 190 countries. Therefore, documents of this kind serve as a solid framework for building strong and human rights-respecting laws.

To begin, let's consider the use of AI during warfare, a situation that is unfortunately prevalent and constitutes a primary concern for diplomats. During war conflicts, AI plays an important role. On the one hand, some tools are used to plan and design war strategies. An example of this is Fire Factory, implemented by the Israeli army to select airstrike targets and process data for various military logistics.

On the other hand, one ongoing concern is "Lethal autonomous weapons," which use AI. These are machines that were designed and created to kill and/or cause damage; they are completely autonomous. Nowadays, there are no strong initiatives to stop or regulate them. In fact, the lack of debate around them and legal vacuum allows States to build and use them without limit.

Autonomous weapons challenge most of the principles of International Humanitarian Law (IHL). The Stop Killer Robots coalition works to ensure human control in the use of force. The campaign calls for new international law on autonomy in weapons systems. It identifies nine problems with killer robots; some of them are: algorithmic biases (explained in Section 4.1), loss of meaningful human control, lack of human judgment and understanding, lack of accountability, and inability to explain what happened or why; which is linked with the huge problem of the black box in AI (explainability and transparency, as we saw also in Section 4.1). According to the International Committee of the Red Cross (ICRC), the International Humanitarian Law principles (present in the international conventions) are: military necessity and humanity; rule of distinction (attacks may only be directed against military objectives not civilians or their objects); prohibition of indiscriminate attacks; prohibition of disproportionate attacks and; obligation to take all feasible precautions in attack. If a soldier has laid down his arms, he/she cannot be attacked. The goal is to neutralize an enemy, not to destroy it.

Taking into account the regulatory framework and AI's main characteristics (explained in paragraph II), we wonder: How will an autonomous weapon be able to distinguish between a soldier and a civilian? How is it going to be proportionate with the attacks? How will AI be able to realize when a soldier is out of combat? Making these kinds of decisions is difficult even for a human. Some agree with the use of lethal autonomous weapons because they protect the lives of people who work in armed forces, and machines do not need to eat, sleep, and have no physical or mental problems (Danesi, 2022).

The main problem with lethal autonomous weapons is not just the lack of regulation, but the lack of debate around the topic. The most advanced global AI regulation is the AI Act, which (as has happened with the General Data Protection Regulation) is supposed to be the compass for AI law as soon as it becomes enforced. Nevertheless, and unfortunately, it excludes the use of AI for military and defense purposes. Let's analyze it taking into account the latest draft agreement.

It clearly states that systems for military, defense, or national security purposes are excluded from the scope of the regulation, regardless of which type of entity is carrying out those activities, such as whether it is a public or private entity. There are many justifications to reach that decision. Article 4.2 of the Treaty on European Union (TEU) and the common Union defense policy covered by Chapter 2 of Title V of the same treaty are subject to public international law. Apart from that, and as regards national security purposes, the exclusion is reasonable for two reasons: the fact that national security remains the sole responsibility of Member States in accordance with Article 4(2) TEU and, by the specific nature and operational needs of national security activities and specific national rules applicable to those activities.

In any case, the final draft clarified that "if an AI system developed, placed on the market, put into service or used for military, defense or national security purposes is used outside those temporarily or permanently for other purposes (for example, civilian or humanitarian purposes, law enforcement or public security purposes), such a system would fall within the scope of this Regulation. In that case, the entity using the system for other than military, defense or national security purposes should ensure compliance of the system with this Regulation, unless the system is already compliant with this Regulation".

Diplomats are not exempt from all those phenomena. They are valuable actors in creating and promoting international agreements to regulate and manage AI and all emerging digital technologies.

4.3. Gen-AI and Its Impact in Disinformation and Democracy

The arrival of ChatGPT at the end of 2022 brought about one important thing: it made AI popular and a subject for light-hearted talk for everyone. That was of great value for all of us who were seeking to raise awareness about the risks of AI and, above all, the construction of an ethical governance framework for AI

Nevertheless, we must acknowledge that ChatGPT introduced new risks in our society. First, as a tool accessible for everyone, everybody can use it but in a good or a bad way. Saying "everybody", it also includes children and teenagers. This is the case of the small town Almendralejo, in Extremadura Spain, where a group of underage youngsters created naked images of their female classmates. This falls short, one of the most dangerous consequences of bad use of GenAI is disinformation and the impact of this arrives to wars.

One instance involved the fake video of Volodymyr Zelensky, Ukrainian's president, where announced that the war was over. Another case was the images shared in X (ex-Twitter) by Prime Minister of Israel, that showed charred babies bodies done by Hamas. Then, some argued that the photos were fake and created by AI. The huge issues with these cases are the following. On one hand, in some cases it is impossible to know whether effectively the content is fake or not. It creates a big doubt and uncertainty, and consequently, a disinformation situation. On the other hand, when the falsehood is proven, it is impossible to communicate it to all the people who have read the fake news.

5. Developing a Practical Guide of AI Ethics Principles in Diplomacy

First of all, we have to affirm that most of the cases listed above are related to ethics concerns but, most importantly, to Human Rights. A clean, healthy and sustainable environment is a Human Right (UN Human Rights Council resolution 48/13); the right to Freedom of expression and information is a Human Right and non-discrimination (for example with an algorithm which makes decisions based on biased variables) is a Human Right. For that reason, we believe that the development and construction of a framework for an Ethics Governance of AI, including AI regulation, has to be a priority for diplomacy.

Consequently, we strongly believe that creating a practical guide of AI Ethic's principles in Diplomacy would be a useful tool for diplomats in order to work as a compass in the new AI world. It will help and guide diplomats in their delicate tasks.

Many documents constitute the foundation of the guide. One of them is the UNESCO Recommendation on the Ethics of Artificial Intelligence and its key tools such as the Readiness Assessment Methodology (RAM) and the Ethical Impact Assessment (EIA). Others important documents are: "Innovation and technological change, and education in the digital age for achieving gender equality and the empowerment of all women and girls" (CSW67 UN Agreed Conclusions), "Feminist contributions in artificial intelligence. Epistemological justice and ethics of equality" (Special representation for feminist foreign policy -Arg-, UNESCO, UN Women and Embassy of the Kingdom of the Netherlands in Argentina), "Artificial Intelligence diplomacy: Artificial Intelligence governance as a new European Union external policy tool" (Policy Department for Economic, Scientific and Quality of Life Policies at the request of the special committee on Artificial Intelligence in a Digital Age), among others.

Considering those documents, the key aspects of the practical guide are the following.

1. **Describing the main tasks of diplomats' work that are linked to or impacted by the use of AI:** this constitutes the introduction of the document and will reflect the state of the art of the relationship between AI and diplomacy. As AI has an exponential growth, the determination of the tasks will be merely illustrative and not exhaustive. The goal is to be clear about the huge impact of AI on the area.
2. **Identifying "red flags" areas taking into account in AI Ethics and Human Rights:** some of them were highlighted in the title IV, such as algorithmic bias, autonomous weapons and GenAI.

3. Creating a list of principles of AI Ethics in Diplomacy:
   (a) *Application of International law and its subareas in AI.* This may sound redundant, but it is not. Nowadays we are experiencing a—justified—fever to reach AI regulation. Although we agree with achieving this objective, it should not be done at any price and even less so, ignoring the current international standards. While we are working on this (the future AI law), at the same time, diplomats also have to work on the reinterpretation of the law in force. We do not need a special rule in order to avoid discrimination through AI in social media. If an algorithm does that, it must be a consequence. The same happens if an algorithm arbitrarily censors content. The validity of international law, reinterpreted in accordance with the Fourth Industrial Revolution, must be vindicated today more than ever.
   (b) *International Cooperation and Collaboration focusing on AI:* Foster international cooperation and collaboration in the development and use of AI in diplomacy to promote shared standards, interoperability, and mutual understanding.
   (c) *Create awareness of AI risks and risk management:* Be vigilant about biases inherent in AI algorithms and data sets must be a priority for diplomats. This has to be pursued not only in diplomatic decision-making systems but also in all of the tasks that they carry out. It is mandatory to actively work to identify and mitigate bias to ensure fair and equitable outcomes. This also includes engaging with the public and stakeholders to foster understanding and awareness of AI in diplomacy, its potential benefits, and associated risks. Finally, as prevention is also a priority, diplomats have to promote the implementation of mandatory algorithmic audits.
   (d) *Data privacy, security and sovereignty:* Ensure robust data security measures are in place to protect diplomatic information and respect national sovereignty in data governance frameworks. Prioritize the protection of human rights and privacy in all AI applications, especially when handling sensitive diplomatic information. Be aware and create awareness that data is not an individual issue, it is a collective interest. When we are giving away our data, we are giving information about us, our family, our neighborhood, our community and our country. It creates our social profile which is used by Big Tech companies and is replicated by algorithms.
   (e) *General principles of AI:* Subsidiarily, diplomats have to follow the general principles of AI such as explainability;

transparency; accountability; human agency and oversight; robustness and safety; privacy and data; diversity, non- discrimination and fairness; societal and environmental; etc.

4. **Designing practical tools to help diplomats' work:** Like this handbook, it could be useful to create practical tools to contribute to diplomats' job. For example, guidelines to create awareness and fight against disinformation such as double- checking the news that we consume (before sharing content on social media), list with reliable sources (officials or international organizations), etc.

6. Conclusions

Through this chapter, we attempt to demonstrate and exemplify the close relationship between AI and diplomacy. We analyze the two main uses of AI in diplomacy: AI as a topic for diplomacy and AI as a tool for diplomacy. Even more important is to be aware of the critical areas in that relation; that is to say, to be clear about which are the principal risks and dangers of AI. Some of these are: black boxes, algorithmic bias, the regulation of AI, autonomous weapons, and GenAI.

One thing is clear: the only way to position AI as an ally in the future construction is to ensure that we are working within an Ethics Governance of AI. Diplomats are an essential element to reach this goal. For that reason, we need countries with clear strategic plans on AI in which foreign policy includes concrete guidelines in the area. We are extremely aware that the practical guide that we are proposing is just the tip of the iceberg that we have to unfold in order to guarantee that no one is going to be left out in the path of AI.

References

Araújo-Moreira, Fernando, Serrano, Nadja, & Migon, Eduardo. (2022 April). From science and technology to innovation diplomacy: their future and the relationship with international security. *Tempo do mundo journal*, RTM (28).

Burkadze, Khatuna. (2021). *Drifting towards digital foreign policy*, vol. 45(2 Summer).

Danesi, Cecilia. (2022a). *El imperio de los algoritmos*. Buenos Aires: Galerna.

Danesi, Cecilia. (2022b). *Struggling against algorithmic bias: Civil liability and other legal remedies.* Bari: Cacucci Editore.

European Parliament. (2021). "Artificial Intelligence diplomacy: Artificial Intelligence governance as a new European Union external policy tool", Policy Department for economic, scientific and quality of life policies at the request of the special committee on artificial intelligence in a digital age.

Fedoroff, Nina. *Science diplomacy in the 21st century.* Elsevier Inc. https://doi.org/10.1016/j.cell.2008.12.030.

Harari, Yuval Noah. (2016). *Homo deus: A brief history of tomorrow.* Harper.

Höne, Katharina E., Anđelković, Katarina, Perućica, Nataša, Saveska, Virdžinija, Hibbard, Lee, & Maciel, Marília. (2019). Mapping the challenges and opportunities of artificial intelligence for the conduct of diplomacy. *Diplo Foundation,* 14.

Leijten, J. (2017). Exploring the future of innovation diplomacy. *European Journal of Futures Research,* 5(20), Springer.

Manor, Ilan & Kampf, Ronit. (2022). Digital nativity and digital diplomacy: Exploring conceptual differences between digital natives and digital immigrants. *Global Policy,* 13(4), 22 May. https://doi.org/10.1111/1758-5899.13095

Mitchell, Tom M. (1997). *Machine learning* (p. 2). Hardcover.

Moreno, A., Armengol, E., Béjar, J., Belanche, L., Cortés, U., Gavaldà. R., Gimeno, J. M., López, B., Martín, M., & Sánchez, M. (1994). *Aprendizaje Automático, Ediciones de la Universidad Politécnica de Cataluña* (p. 208). Barcelona.

Qian Xia, Lucie. (2024). Diplomatic relationship-building in the age of generative AI: the European Union and China. *Place Branding and Public Diplomacy.*

Robyn, Williams & Otto, Lisa. (2022). Artificial intelligence as a tool of public diplomacy: communication between the United States and Iran. *The Thinker,* 90.

Shannon, Claude. (1938). A symbolic analysis of relay and switching circuits. *Transactions American Institute of Electrical Engineers,* 57, 713–723.

Shannon, Claude, Wyner, A. D., & Sloane, N. *Claude E. Shannon: Collected papers* (1st ed.). USA: IEE Press.

Special representation for feminist foreign policy, Feminist contributions in artificial intelligence. Epistemological justice and ethics of equality, 2024.

Stanzel, Volker & Voelsen, Daniel. (2022). Diplomacy and artificial intelligence: Reflections on practical assistance for diplomatic negotiations. SWP research paper, no. 1/2022, Berlin.

Torres Jarrín, Mario. (2021). *EU & Ethics governance of the artificial intelligence: Artificial intelligence & diplomacy.* Cuadernos salmantinos de filosofía.

Turing Alan M. (1950). *Revista mind, computing machinery and intelligence.*

UN. (2023). Innovation and technological change, and education in the digital age for achieving gender equality and the empowerment of all women and girls. CSW67 Agreed Conclusions.

UNESCO. (2021). Recommendation on the ethics of artificial intelligence.

UNESCO. (2023). Ethical Impact Assessment (EIA).

UNESCO. (2023). Readiness Assessment Methodology (RAM).

US. (2023). *Executive order on the safe, secure, and trustworthy development and use of artificial intelligence.* The White House.

Véliz, Carissa. (2021). *Privacy is power: Why and how you should take back control of your data.* Bantam Press.

Chapter 3

Science Diplomacy and Innovation Diplomacy: European Trajectories in Information and Communication Technologies (a Long-Term Perspective)

Pascal Griset

Abstract: This article, anchored in a historical approach, will highlight what an appropriation of innovation diplomacy by the European Union would allow to diversify and strengthen its influence. It will also highlight the risks that this new form of diplomacy, at the crossroads of science, technology and economics, must take into account. Risks that are not only diversified but also systemic when it comes to integrating the threats generated by what is known as cyberwarfare. In this respect, the implementation of such a diplomacy could not be done in an "isolated" manner. To be fruitful for the EU, it would indeed imply very significant changes in the global mode of operation of the Union as well as a reconsideration of its relationship to the notion of sovereignty.

Keywords: Diplomacy. Information technologies. Communication technologies. Cyberwarfare. Sovereignty. European Union.

1. Introduction

The concept of science diplomacy, undeniably prolific, seems to have been applied to multiple practices at the risk of becoming a portmanteau word, a comfortable vector whose multiple interpretations have favored the most diverse projections and facilitated the expression of a universalist thought with more or less consciously performative ambitions: science and knowledge will lead humanity to a better future. The proximity in daily action of various domains, structurally dependent on each other, has also favored this evolution. Thus, the interweaving of science and technology, characterized by the concept of "technoscience"[1], has led to implicit

[1] Pestre, Dominique (with Yves Cohen). (1998). "Histoire des techniques", *Annales, Histoire, Sciences Sociales* 53(4–5), 721–744.

or explicit[2] association of techniques with science-related diplomatic practices. In the same way, the connections inherent to scientific practices (economic and financial stakes) or diplomatic practices (intersection with cultural diplomacy through certain actors such as universities) seem to require a redefinition of the limits of science diplomacy. This is done through initiatives that attempt to reconsider diplomatic practices through specific fields of action (environment diplomacy, health diplomacy, energy diplomacy, etc.). These approaches, pragmatic and useful, do not however avoid (one could even think that they enhance it) the need to better define the perimeter and the specific modalities of action of science diplomacy as an enhancement of the work already done in this perspective. In this respect, the work carried out in the framework of the Horizon 2020 programs constitutes a very complete reference available online.[3]

Our proposal, in order to make a limited but hopefully useful contribution to this process, will be based on the concept of innovation diplomacy[4], which has been developed over the last few years and, we believe, is particularly well suited to re-examining the fields of action concerned[5]. In this volume, we will limit our approach to the field of information and communication technologies within the European space around three questions: How did innovation diplomacy emerge in this specific field at the turn of the 19th and 20th centuries in a dynamic that saw Europe progressively dispossessed of its leadership by the United States, in a field that Europeans controlled more or less collectively since the 1850s? We will then examine how this new deal gave rise to new dynamics in which the European states realistically became involved. Finally, we will wonder how the European Community and then the Union, despite their initiatives, were confronted with great difficulties to really deploy an effective innovation diplomacy.

---

2 Flink, Tim and Ulrich Schreiterer. (2010). "Science Diplomacy at the Intersection of S&T Policies and Foreign Affairs: Toward a Typology of National Approaches", *Science and Public Policy*, 37(9).

3 https://www.science-diplomacy.eu/science-diplomacy-resources/

4 Bound, Kirsten. (2016). "Innovating Together? The Age of Innovation Diplomacy", in: Soumitra Dutta, Bruno Lanvin and Sacha Wunsch-Vincent (eds.), *Cornell University, INSEAD, and WIPO: The Global Innovation Index 2016: Winning with Global Innovation.* Geneva: World Intellectual Property Organization, pp. 91–96.

5 Griset, Pascal. (2020). "Innovation Diplomacy: A New Concept for Ancient Practices?", *The Hague Journal of Diplomacy*, 15(3), Online 19 August 2020, 383–397.

2. International Telegraphy: From Technical Diplomacy to Innovation Diplomacy

2.1. *Electric Telegraphy: International Organizations and Techno-Diplomacy*

From the 1840s onwards, Europe set up an electric telegraphy network centered on national capitals. These networks were operated by administrations within the framework of a state-owned monopoly. Only the United Kingdom was an exception to this model for a few years. The pioneering companies that were created there at the turn of the 1830s and 1840s, directly linked to the development of the railway network, developed the telegraph network in a very dynamic manner for more than ten years.

The inequality of service between large and small cities, however, placed the private companies under increasing criticism. Telegraph had become a necessity and was consequently considered as basic infrastructure[6].

This led Parliament to vote on the 1868 Telegraph Act, allowing the purchase of private lines by the Post Office. The nationalization of the telegraph was completed in 1870. Thus, by the 1870s, there was a uniform system of public monopoly over electric telegraphy. It created a homogeneous space with regard to a political economy that placed the State at the heart of the system. In the United States, the evolution was radically different. The development of the telegraph network was part of an extreme competition, which saw, within a politically and linguistically unified space, a multitude of private companies confront each other. Through bankruptcies, takeovers and mergers, a private duopoly emerged in the last third of the 19th century around two large private companies, the Postal Telegraph Company and the Western Union. Europe and North America thus opted for two opposing models in terms of the political economy of telecommunications. The construction of the European telecommunications space, starting from a politically fragmented and often conflicting territory, nevertheless required innovation in terms of international cooperation. This experience was in many

---

6 Cohen, Ira T. (1980). "Towards a Theory of State Intervention: The Nationalization of the British Telegraphs." *Social Science History,* 4(2), 155–205.

respects the foundation for the creation of a model for international telecommunications governance.

The first international institutions were set up in Europe in the middle of the 19th century. The Berlin-Vienna telegraph link project at the end of the 1840s led to the signing of the first international telecommunications agreement between Prussia and Austria in 1849. The following year, the creation of the Austro-German Telegraphic Union initiated a movement that gradually involved the majority of the powers concerned. The Convention between Belgium, France and Prussia in 1852, the Berlin Conference (1855) and the Brussels Conference (1858) were all initiatives that led to the signing in Paris on May 17, 1865, of the first international telegraph convention creating the International Telegraph Union. The exceptional nature of this development is underlined by the fact that the Universal Postal Union, which was created nine years later, was managing an older technique. Three conferences, Vienna (1868), Rome (1871) and St. Petersburg (1875), established common procedures, standards and tariff regulations that provided an indispensable framework for the growth of international connections[7].

The construction of these institutions can be considered as a founding period setting in motion a technical diplomacy, carried by administrations[8]. If science was not absent from the debates, it even constituted an important substratum, the real stakes were commercial and political and were part of a dynamic of cooperation. This led to the creation of permanent institutions that managed to develop in a Europe marked by the Crimean War, by the conflicts between Prussia and Austria, and then between Prussia and France.

This period was marked by the need to connect the wire networks of different European countries[9]. The connection between these networks and the American telegraph network, which became a reality in the second half of the 1860s with the break of the first transatlantic

7 Fickers, Andreas and Gabriele Balbi. (2020). *History of the International Telecommunication Union; Transnational Techno-Diplomacy from the Telegraph to the Internet.* Berlin: De Gruyter.

8 Laborie, Léonard. (2008). "En chair et en normes: Les participants aux conférences de l'Union internationale des télécommunications, de sa fondation à sa refondation (1865–1947)'. *Flux,* 74(4), 92–98.

9 Fickers, Andreas and Pascal Griset. (2019). *Communicating Europe.* London: Palgrave Macmillan.

cables, had been made possible, in terms of political economy, by the use of private initiative. The cables manufactured in Europe and laid by European operators (whether English, French or German) were therefore operated by private companies, benefiting from a concession granted by their respective governments, which placed them on the bangs of the State's monopoly on the electric telegraph. Since the United States was not willing to join the International Telegraph Union, a compromise had to be found to allow two networks, managed on completely different legal bases, to connect. It was finally on the basis of American rules that an international consensus was reached that allowed the interconnection of these two spheres ... The American space, regulated by the market, thus communicated with a European space composed of multiple sovereign states and regulated by the complex games of technical diplomacy, on the bases that it had itself set. Consequently, the British, French and German governments conceded the operation of transatlantic routes to private companies over time. These companies remained under the close control of the administration within the framework of a state-owned monopoly on the telegraph which remained the rule.

2.2. Wireless Telegraphy and the Beginnings of an Innovation Diplomacy

Wireless telegraphy, developed by many companies on the basis of the innovation made by Guglielmo Marconi in 1896, radically changed the game at the turn of the 19th and 20th centuries. The development of this new field, which was primarily dedicated to naval communications and then to long-distance international services, became globalized very early on. Moreover, the multiple transmitters deployed throughout the world by a wide variety of actors, free to choose the technical devices and wavelengths that suited them, made it necessary to set up coordination procedures to avoid interference problems and the "chaos in the air" feared by many specialists. Finally, wireless telegraphy quickly became different from wire telegraphy because of its extreme technical flexibility. Indeed, while wire telegraphy had stabilized quite quickly technologically, with few notable developments in more than half a century, wireless links were marked by a veritable bubbling of innovations, constantly calling into question the positions of some and others.

However, Marconi had tried to establish this stability by relying on his patents and on an international development carried out on both sides of the Atlantic. Founded in London in 1897, his company set up an American subsidiary in 1899, the Marconi Wireless Telegraph Company of America. The group quickly took a dominant position on the European and American markets. By 1903, its system had 45 coastal stations and counted many shipping companies among its customers. In Germany, the many initiatives taken in response to Marconi's initiative were supported by the state at a very early stage. However, in order to prevent the efforts of the Reich's companies from being dispersed, the government quickly encouraged the two large groups in the electrical industry, AEG and Siemens, to join forces. The result was the creation in 1903 of the Gesellschaft Für Drathlose Telegraphie (Telefunken), which was created by pooling the radio divisions of the two electrical companies. Rathenau, Bredow and the Kaiser himself were the most prominent shareholders in a company that was clearly a national champion, operating in Central Europe, Sweden and Russia. An American subsidiary, the Atlantic Communications Company, was created in 1905. The company's entrepreneurial ambitions were therefore very quickly international, and even global, since Telefunken was also very quickly involved in the North and South American markets. If the British government was somewhat reluctant to fully support the development of Marconi, created by an Italian ... it was also very clear that the ambitions of the two companies crossed those of the two empires.

To consolidate its initial advantage, Marconi Wireless tried to neutralize, or at least to limit, the development of competition by trying to impose on its customers the prohibition of communicating with the stations of other companies[10].

It justified this exclusivity clause by asserting, against all technical logic, that its stations could not ensure a quality service by communicating with equipment that was not strictly identical to its own. In fact, Marconi, on the basis of its very dominant control of the world market, was trying to establish a real monopoly. This project, known as "Marconism", provoked a very strong international reaction. The fact that this claim came from a British company, linked to the same Power that already controlled most of the trans-oceanic cables, simply increased the concern. The military and political implications of radio were such that the debate could not remain limited to the strict economic domain.

---

10 Except in case of distress ...

Germany, for very clear strategic reasons in the context that would lead to the Franco-British rapprochement of 1904, decided to launch a counterattack.

In 1903, it chose the "weapon" of international consultation by proposing to the heads of the main administrations to meet in Berlin to study the problems related to the organization of international radio communications. In fact, the "[...] only real reason for holding this conference was to determine whether all the countries or companies could act in the way Marconi had decided to act". With the exception of England (which supported Marconi wireless as a British company) and Italy ("loyal" to the Italian Guglielmo Marconi), on August 13, 1903, all the delegations signed a memorandum of understanding affirming freedom of communication in the face of the hegemonic claims of the British industrialist, and stressed the principle of freedom of communication, specifying that the coastal stations and the on-board stations had to exchange their messages "without distinction of the systems of communication. . without distinction of the systems of wireless telegraphy employed". This event marks the beginning of a new crucial phase of development for techno-diplomacy linked to a very intense evolution of international organizations.

2.3. *Competition and Negociations: Diplomacy in Practice*

The first International Radiotelegraph Conference belonged to this dynamic. Meeting in Berlin in 1906, it still had to deal with a few questions linked to the failure of Marconi who tried to obtain some compensation, but it concentrated above all on the indispensable work required to give to the development of the new technology a better appropriate framework. International institutions had been created for classical telegraphy to allow the connection of networks. In radio, the objective was more complex and involved a greater number of variables. First of all, it was necessary to limit mutual jamming between transmitters that required a lot of radio spectrum. A first allocation of frequencies, the adoption of procedures regulating communications as well as a first technical standardization constituted the essential achievements of the conference. A first text settled the framework of a radiotelegraphic convention, a second one establishing a "Radio Rules". They came into force in July 1908. Unlike the creation of the Telegraphic Union, private companies were involved in this process, just as they played a major role in the development of the networks. National diplomacy, starting with the Reich, supported the companies in an unprecedented way. Telefunken, supported by German financial circles and subsidized by the government, relied constantly on Reich diplomacy to promote its international activities. The latter, which had played a decisive role in the failure of Marconism, did not relax its efforts, opposing with increasing success Marconi's efforts to penetrate many countries. This behavior of the German company, described by Marconi Wireless as the "Telefunken wall"[11], The British company initiated lawsuits against its German competitor in order to enforce the rights it held. The tension was so great that the Reich government decided to ban all foreign equipment on German ships (the only ones concerned were in fact Marconi equipment…). Negotiation quickly followed this confrontation phase and led to the signing of an agreement between Marconi Wireless, the Brussels Wireless Telegraph Company (licensee of Marconi patents for Germany and several other European countries) and Telefunken. It paved the way for the signing of a cross-licensing agreement between the two companies in 1912 and the establishment of an informal market sharing between the two groups. This very strong combination between the diplomatic networks and the "champions" of the two powers presented a totally different situation from that which prevailed until the 1890s.

This conflictual convergence of efforts to promote national interests, combined with a sense of negotiation and compromise, allowed the emergence of two European companies of worldwide stature. Both companies were born and grew in countries that had instituted a state monopoly on the telegraph and telephone. The different traditions in the application of this monopoly had, however, provided a very different framework for development. German centralism was opposed by British liberalism. Both philosophies were, however, applied with strong nuances which may suggest a certain paradox. The British "liberalism" allowed the Post Office to intervene to a large extent, while the German "centralism" relied on private enterprise to initiate the new technology.

Both experiences showed that the development of international radio links required a balance between different logics and interests and a

---

11 Baker W. (1971). *A History of the Marconi Company.* New York, pp. 131, 414 p.

pragmatic adaptation to the international environment. The schemes applied until then for national networks were inoperative.

The development of radio rapidly raised radically new problems in terms of international cooperation. Its evolution was characterized by the existence of two contradictory trends. While the technical nature of radio made the setting up of international bodies indispensable in order to combat interference, its economic structure, which left a very important place to private initiative, tended to keep it away from existing organizations, since they were designed for and dominated by administrations. It could therefore only be, at first, within a specific framework that radio could establish rules shared by the different nations.

Consequently, the techno-diplomacy carried out within the International Telegraph Union, for wired telegraphy by administrations was superseded by an innovation diplomacy customized for wireless telegraphy by a by a hybrid community of actors where public administrations and private companies meet, not without difficulties. This new community of actors was all the more difficult to stabilize because their technical cultures were not uniform. The new diplomatic space was distinguished by the major place of companies and by the crucial nature of good cooperation between the companies of a country and their national diplomatic representation. A new diplomatic space was thus established in the 20th century. With regard to the practices of the previous one, it had to integrate the almost structural instability generated by the innovative activity of companies adopting the world as a new horizon

3. Private Companies at the Heart of Innovation Diplomacy

3.1. American Practices: New Scales, New Actors

While the United States had never wanted to apply for membership in the International Telegraph Union, it did become directly involved in the establishment of international radio institutions. As early as the 1903 preliminary conference, it seemed clear that the United States would take an active part in organizing the new networks. In 1904, President Roosevelt created an interdepartmental commission of five members. It was charged with defining an American position for the next international conference announced on radiotelegraphy issues. It worked by associating to its deliberations the companies involved in the field. Their opinions were very mixed, but the process that was put in place allowed the definition of a shared position that gave the American delegation a more legitimate influence.

The settlement of the First World War clearly shows that this new paradigm had become established. A complex process, carried out on the bangs of the Versailles negotiations (while the question of submarine cables was dealt with directly by the delegations of the Peace Conference) had enabled the Allies to prepare a reorganization of the collective management of international telecommunications, leading mainly to the merging of the procedures concerning submarine cables and radio[12]. This project, formalized by the delegations of the victorious countries in a memorandum dated August 25, 1919, was to be confirmed at a conference organized in Washington on October 8, 1920. The delegations, composed of military personnel and representatives of the administrations concerned, hoped to conclude the agreement quickly. They had apparently forgotten that the text had been drafted without any real consultation with the private companies. This negligence led to a rude awakening. A communiqué, drafted by the major U.S. companies a few weeks before the conference, strongly criticized large parts of the memorandum. In particular, the hypothesis of a merger of the two unions was totally rejected[13].

Forced into a meeting with representatives of the big companies, the official American delegation did not manage to find a compromise and the conference produced a draft international convention that planned to merge the telegraph and radio unions. This forced passage was met with resistance from private companies. Opposition from companies ranging from cable companies to RCA and AT&T only grew stronger. Although less virulent, because they were probably more used to dealing with administrations that based their power on the existence of state-owned monopolies, European companies followed in the footsteps

---

12 Pascal Griset. (2023). "The Treaty of Versailles and Intercontinental Telecommunications: Between the End of the War and the New Techno-Strategic Order, 1917–1919", in: Laurence Badel (dir), *The Paris peace Conference, 1919.* New York/Oxford, Berghahn, to be published.

13 E.U-F-GB-I Radio Protocol of August 25, 1919, as modified and commented upon by a Committee Appointed by the Secretary of Commerce, Washington DC, GPO, 1920.

of their American counterparts. It is true that they were indispensable partners and that they were already bound by numerous contracts. It is consequently a transnational consensus that links private companies to form a group to lobby their respective governments and international organizations. The resulting blockage was total and made obsolete the project. The merger of the two fields was postponed for several years...

This capacity of private companies to influence the decisions of an international conference and then to block an evolution that seemed opposite to their interests highlights the passage from a technical diplomacy, managed by the civil servants and engineers of the State-owned networks, in the continuum of a shared culture, to an innovation diplomacy in which the companies and their rationale take a decisive part.

This highlighting of the new role of the actors allows us to corollary propose the idea that the characterization of innovation diplomacy is possible by identifying the concerned parties and their weight in the process of negotiation and then of decision. It allows us to see that the place taken by private companies, in a transnational solidarity between Americans, Germans, British and French, is the key indicator of this innovation diplomacy. The failure of the Washington Conference in 1919 also suggests that the ability of diplomacy to ensure fluid dialogue, conducted at a very early stage with the private sector of its country, is the prerequisite for serving the national interest, in which companies consider themselves to be key players.

Analyzing the confrontations between international companies, a US Navy analyst stated in 1919: "There is at present no competitor who can hope to win without the help of his government. Real collaboration between private enterprise and the Administration could be a major asset for the United States. If our government can coordinate the activities of our private companies with those of the Administration, we shall be in a much better position than France and Great Britain, which are moving toward single state control. These governments will see their action limited to their own territory. The hostility existing in some European countries between private and public radio systems is therefore totally to our advantage, if we have enough foresight and if we show ourselves capable of taking advantage of it."[14]. The Army largely shared this analysis. The Signal Corps regretted very deeply the difficulties of building a

---

14 Note du Department of State to Secretary of the Navy, 4 October 1922, Naval Records Collection of the Office of Naval Records and Library, National Archives, RG 45, Box 983.

consensus between the private companies and the U.S. Administration that would allow the development of a true nationally based foreign policy. "The attitude of our telecommunication companies has often been negative toward international conferences, based on the fact that the official U.S. delegations have not represented their interests in the desirable manner. These companies are hostile to conferences because they believe that their interests would be much better served if they were free to arrogate to themselves a squatter's right to as much radio space in the world as possible. It seems unlikely that such a policy will pay off in the long run and it would seem far better to determine a real policy and work for its success."[15]

Combined with the creation of Radio Corporation of America, these developments gave the United States a force unknown until then. They were also integrated into a desire to export an American model based on economic liberalism and parliamentary democracy. A moral dimension could not be excluded. As its president declared in 1921: "RCA is one hundred per cent American [...] its growing success is an indication of American ideals [...] its affairs will always be conducted according to the fine ideals of service for country and for humanity [...]".

3.2. A European Innovation Diplomacy in South America against the Wireless Monroe Doctrine

Until the First World War, only two radio companies were significantly interested in the South American market. They were European: Marconi and Telefunken. This situation did not suit the US administration. In Washington, in January 1916, during the Second Pan-American Scientific Congress, the State Department organized an "informal" meeting to present to the other nations its positions on the development of radio links in the continent. 19 countries responded to the invitation[16]. The

---

15 US Congress, Senate, Committee on Interstate Commerce, 66th Congress, A *Bill to Prevent the Unauthorized Landing of Submarine Cables in the United States: Hearings before a Subcommittee of the Committee on Interstate Commerce on S. 4301.* 3d sess. (15th December 1920–11th January 1921).

16 Guatemala, El Salvador, Honduras, Costa Rica, Nicaragua, Panama, Colombia, Ecuador, Venezuela, Peru, Bolivia, Chile, Argentina, Uruguay, Paraguay, Brazil, Cuba, Santo Domingo and Haiti. Minutes of the informal conference on radio communication, held at the Departement of State. 7 Jan. 1916, p. 1.

message of the State Department was without ambiguity. After having expressed his interest in the scientific dimension of radio waves, the speaker regretted the importance of European industries in the construction of radio stations in Latin America, and worried about the risk of seeing the neutrality of the countries considered called into question by the presence on their territory of stations under the indirect control of conflicting countries. He expressed the State Department's conviction that this vital technology "[...] should not escape this hemisphere and fall under non-American jurisdiction". Josephus Daniels, Secretary of the Navy, proposed a strategy that would allow Latin American nations to defend "the principle of America for Americans". This policy soon proved to be a complete failure for Washington, as American industry proved unable to offer Latin American governments a credible alternative to the British and German proposals. The government's hope of enforcing a "Wireless Monroe Doctrine", which would exclude the participation of European companies, whether or not they were subsidized by their governments could not materialize.[17]."

Events went in a completely different direction because Telefunken and the Compagnie Générale de Télégraphie sans fil were able to obtain concessions from several South American governments to build their own intercontinental radiotelegraph station. While Marconi believed that it would be ruinous to try to compete with these powerful companies, given the investments and the volume of traffic expected, the idea of an agreement between Europeans appeared to be very realistic. The Americans were very reluctant at first, but finally rallied to the idea of joining this three-party agreement to make it a four-party structure, which was not very compatible with the "wireless Monroe doctrine" but which allowed them to remain in the game. On the basis, unchallenged by the Europeans, that the United States should retain a leading role in any partnership based on its fundamental strategic interests in South America, a balanced agreement was finally negotiated. The treaty was ultimately signed on September 19, 1921. The four companies pooled all their concessions held in South America. A "Centralization and Coordination Committee[18]", which was based in Paris, had the task of

17 Admiral Bullard's Report to the RCA Board of Directors, sept. 1920. Archives AT&T, Box 50.

18 Girardeau É., *Souvenirs de longue vie*, p. 132. The American wording is "Board of trustees".

developing for the common interests of the partners the radiotelegraphic links between South America and the other continents. It was composed of eight members, two for each company, plus a president, of US nationality. The agreement was greatly facilitated by the context of technological uncertainty. None of the companies could base their future on a technology that would allow them to go it alone and attempt to compete on tariffs and trade.

The analysis by the American Administration of what was called in Washington the "AEFG trust[19]", deserves all our attention to evaluate the meaning of this agreement. This text was indeed considered a defeat by the US Navy. In spite of the privileged status reserved for American interests within the board of directors, some, and especially Commander Hooper, who had played a key role in the constitution of the RCA, felt that the ambition of the Wireless Monroe Doctrine had been betrayed. The man who, in the shadow of Admiral Bullard, had been the main inspirer of a nationalist policy wrote all his bitterness in the spring of 1922: "My disillusionment is such that it has turned a young man of patriotic enthusiasm into a sad and disappointed old man.[20]"

4. Innovation Diplomacy: A Vector for Unifying and Strengthening the European IT Industry?

The dynamics that allowed European networks to develop technically in a coherent way were almost naturally integrated into the actions of the European Union. The European customer, beginning with all European organizations, was placed at the heart of a dynamic market where several major players offer all the necessary services. The "European" business was not at a disadvantage compared to the conditions offered to its American or Japanese counterpart. Since the RACE program (1987), cooperation on infrastructures with the aim of homogenizing the equipment of the European territory has been[21] a major concern for the Union. The new members have thus benefited from extensive

19 For America, England, France, Germany.

20 3 June 1922 letter to O.D. Young, Hooper papers, Library of Congress, Box 4.

21 *Research on Advanced Communications in Europe. Vers les télécoms de l'an 2000: lancement du programme RACE*, Communautés européenne. Commission, Luxembourg, Office des publications officielles des Communautés européennes, 1988, 17 f.

aid to bring their networks up to international standards as quickly as possible so that their integration would not be jeopardized by a possible "digital divide". The Union is also particularly vigilant in ensuring that this market is truly open. The implementation of mobile networks has demonstrated its ability to standardize (GSM[22]) and then to gradually impose tariffs that were less unfavorable to users.

However, another challenge arose for the Community and then for the European Union. It was about making Europe a coherent space for innovation in order to allow European information and communication technology companies to play their full role in the face of the very strong competition coming from the United States, then from Japan, then from China. To do this, it was necessary to get European entrepreneurs to work together. This convergence of the initiatives, stimulated by the actions of Brussels, had to both strengthen the performance of the companies and homogenize the actions of the different member states.

4.1. Joining Forces to Be Stronger: The Narrow Gate

A first type of scenario could be envisaged in the European space. Innovation diplomacy can be carried out by two states wishing, on their own initiative, to promote a convergence of their policies and encourage a partnership between the companies of their respective countries. The deregulation of telecommunications, which had seen a major withdrawal of the State in the field of information and communication, constituted an important stimulus in this respect. The policy of liberalization of telecommunications carried out since the 1980s had indeed confronted the national operators with an immense challenge. Heirs of the PTT administrations and known as "historical operators", they were only able to leave their own premises at the cost of extremely costly investments, made at the worst moment of an Internet bubble that their emergency strategy had helped to build. The idea of bringing together the main players in order to create European companies that could reach

---

22 Global System for Mobile communications. Dudouet, F. X., Mercier, D., and Vion, A. (2006). « Politiques internationales de normalisation, Quelques jalons pour la recherche empirique », *Revue française de sciences politiques*, 56(3), 367–392; Laborie L. (2008). « Concurrence et changement technique. De la norme au marché, la trajectoire "unique" de la téléphonie mobile en Europe depuis les années 1980 », *Histoire Economie et Société*, 1, 91–101.

critical mass on a global scale emerged and constituted a way of innovation diplomacy between some member states. France and Germany thus tried to encourage a closer relationship between their two historical operators. This project had all the characteristics of an ambitious "innovation diplomacy". It was supported by the diplomatic networks of both countries and concerned a high-tech field, marked by major developments around digitization, the development of Internet and mobile uses. At the end of the 1990s, mainly at the instigation of the French government, a merger between France Telecom and Deutsche Telekom was envisaged. Everything seemed to be in place to build an "Airbus of telecoms". Two administrations of comparable weight, having followed with great caution the path towards the liberalization of the European market. Two communities of engineers and administrators experienced in the mysteries of "virtuous" cooperation between network operators and national industry. These similarities were, however, very superficial. Without going into the fundamental differences rooted in the history of these two entities, it seemed somewhat illusory to ask companies that were now private to get along in a climate of "goodwill" when everything in the community's policy encouraged them to confront each other. However, the political imperative and the diplomatic relations between the two countries pushed this idea. In the name of the Franco-German "couple" the banns were published. In July 1998 a strategic agreement was signed between the two companies. It is based on an exchange of shares, admittedly at a very symbolic level. The signatories of this agreement seem to want to rely on pre-existing synergies residing mainly in partnership agreements such as the joint venture between France Telecom, Deutsche Telekom and Sprint in 1996 or the agreement signed between Wind, France Telecom, Deutsche Telekom and ENEL for Italy in November 1997. This friendship will be particularly emphasized by France Telecom, anxious to please the French government in a turbulent context. The French company sees strategic "concertation" on both sides of the Rhine as a key element in its operations, and the alliance with Deutsche Telekom is even presented as the "red wire" of its own strategy. This vision would collapse in a drama comparable—on a smaller scale—to that which took place during the UNIDATA affair twenty-five years earlier.[23]. The "betrayal" this time seems to be German, with Deutsche

---

[23] Griset, P. (ed.) (1998). *Informatique, politique industrielle, Europe: entre Plan Calcul et Unidata*. Paris: Institut d'histoire de l'industrie, Editions Rive Droite, 588 p.

Telekom unilaterally deciding in April 1999 to embark on a merger project with Telecom Italia, leaving its companion on the side of the road[24]. Each will then take his destiny in hand and will lose in the tumult of the explosion of the Internet bubble a good part of his assets[25].

This episode belonged clearly to a plan that went beyond the simple framework of an alliance between two firms. The "diplomatic" dimension of the affair was very clear in the commitment of the two States to a field directly connected to political initiatives and to sovereignty issues. Taking into account the control of high technologies, directly linked to industrial and financial issues, allows us to understand the heart of what is "innovation diplomacy". The unfortunate end of the operation also underlines the fact that the German and French governments had probably not taken into account what should be an effective action in this field. The involvement of the companies concerned, on whom the project had been more or less directly imposed, thus proved to be somewhat fragile... The lessons of the UNIDATA project had apparently not been taken into account.

A second type of scenario takes place in another dimension when the European authorities take certain initiatives to encourage projects intended to bring together companies from different member states. The Galileo project is a typical example in this respect. It underlines all the difficulties that Europe faces when it comes to creating a real dynamic of innovation between member states around a strategic imperative[26]. As in the previous example, the political initiative crosses high technology, financial interests and international issues directly linked to sovereignty. Since the end of the 1990s, Galileo has been part of a rationale to catch up with Europe in the field of positioning; Europe must free itself from the American GPS monopoly. Since 1999, a series of selections between different concepts has been organized.

---

24 The article by Claude Soula in the *Nouvel Observateur* of April 29, 1999 is entitled « Le mariage qui fait pleurer France Telecom »,. See also *International Herald Tribune* April 17, 1999: « A european phone giant in works? Telecom Italia talking with Deutsche Telekom », and April 23 *The Tech on line Edition* mentions « The biggest corporate marriage in history » (vol. 119, Issue 21).

25 Plane, M. (2004). « Le secteur des télécommunications surfe-t-il de bulle en bulle ? », *Revue de l'OFCE*, n° 88(1), 248 p., pp. 151–184.

26 Autret, F. (2007). « Quelle organisation pour l'Europe spatiale ? » *Politique étrangère*, (2), 281–292.; Nardon, L. (2007). « Où va le programme spatial français ? », *Politique étrangère*, (2), 293–305.

The main European companies in the sector joined forces in competing groups. Very quickly a series of pitfalls appeared. Rivalries between companies that one would like to see cooperate, but which are incited by the global policy of the Union to confront each other; the notion of "fair return" on the part of the project financed by the Union... the small accounts and short-term calculations will dampen the initial enthusiasm. A first success, somewhat theoretical but nevertheless essential, is however registered with the signature in May 2003 of an agreement between the Union and the European Space Agency. This document foresees then that the public financing could not exceed one third of the investments, the two remaining thirds having to be mobilized by the private partners. Although "strategic" this project cannot indeed constitute a too shocking exception to the liberal principles dominating in Brussels... The process gets bogged down. In 2005 two groups remain in the running, the European authorities, unable to decide, finally ask the two projects to reach an agreement ... While the United States, using the recipes used in the 1970s to undermine the European space policy, stir up behind the scenes the possibility of a more global agreement on the use of GPS, granting more guarantees to the Europeans, the discussions stagnate. It was not until 2007 that the evidence of public funding for a project whose main objective was to ensure Europe's independence in a sector that had become vital for all its activities finally became clear. An "equitable" arrangement between six lots was established with the objective of opening in ... 2013[27]. In 2022, only a part of the satellites was in orbit and the complete deployment of the system remained to be achieved ... a "success" that would take a long time to take shape ....

Through these few scenarios, we can see the necessity, but also the difficulty of building an internal "innovation diplomacy" allowing European Union member states to agree in order to face global challenges to their security. Classically, the importance of the stakes, which could encourage States to overcome their differences, exacerbates on the contrary their desire to control what they consider to be their sovereignty.

27 http://ec.europa.eu/transport/galileo/index_en.htm
http://www.esa.int/esaNA/index.html

4.2. Connecting Science Diplomacy and Innovation Diplomacy: The Smartest Way ? ...

Scientific diplomacy, by setting aside the most immediate economic stakes, can make it possible to initiate rapprochements, favorable to the development of innovation, in a way that is nevertheless likely to minimize antagonisms.

The convergence between diplomacy, science and innovation is particularly striking in the EUREKA project launched in 1985 at François Mitterrand's initiative[28]. From 1985 onwards, this project concerned a fairly broad set of technologies, but information and communication technologies took on an essential role, either specifically or in conjunction with other sectors. This program was intended to be a reply to the proposals made to the Europeans by Ronald Reagan regarding their potential involvement in the Strategic Defense Initiative. Confronted with what he considered a threat, François Mitterrand took proactive and political action to meet the "technological challenge" and avoid a disorderly rallying of Europeans to the American program. The perspectives were presented in a very broad manner. The aim was to achieve better coordination of theoretical and applied research. The project of a University of Europe and of a European Academy of Sciences and Technology was advanced in this perspective. The challenge in the eyes of François Mitterrand was not only economic but: "... intellectual, even spiritual". During the Eureka intergovernmental conference held in Hanover on November 5, 1985, the main elements of the project were put in place. They integrated a logic of catching up in order to place Europe in a stronger position within the Triad. Without neglecting its effects in terms of stimulating international cooperation, the political and communication dimension of the EUREKA programme was obviously the most significant.[29]. This is most clearly reflected in the speech given in Hanover by Robert Goebbels, Luxembourg's Secretary of State

---

28 Braillard, Demant A. (1991). *EUREKA et l'Europe technologique.* Bruylant, Bruxelles, Paris, 174p.; Saunier G. (2006). « Eurêka: un projet industriel pour l'Europe, une réponse à un défi stratégique », *Journal of European Integration History*, 12(2), 57–74.

29 On F. Mitterrand's policy regarding Eureka, see the testimony of J-P Karsenty « Du CESTA à la création d'Eureka » : http://www.mitterrand.org/Du-CESTA-a-la-creation-d-EUREKA.html and « 20 ans de soutien à l'innovation européenne », Secrétariat Eureka, Septembre 2005, 67 p, p. 12 : http://www.eureka.be/files/:873478

for Foreign Affairs, who saw Eureka as "... a political signal to the outside world and a public demonstration of the political will to unite all efforts and bring together all capabilities to meet the great challenge of Europe's technological renewal."[30]. A strategic logic of catching up was indeed at the heart of the diagnosis. "Faced with the delay accumulated by Europe in the mastery and exploitation of advanced technologies, faced with the threat of seeing the technological gap gradually become a real threat to the economic and social progress of our countries, the union of all our forces is more necessary than ever. However, the political dimension, with diplomatic consequences, was at the heart of the approach. As Goebbels emphasized, it will indeed be necessary to succeed: "... to overcome national borders, which are often mental borders.[31].

Eureka did not revolutionize European cooperation, but it was undeniably a political turning point. The American offensive was blocked and a new impetus was given to the actions already launched to promote the establishment of a European Research Area.

The dynamic continued with the successive Framework Programmes for Research and Technological Development (prepared since 1983), which aimed to structure and sustain this momentum over time.

The foundation in 1989 of the European Consortium for Informatics and Mathematics by GMD for Germany, INRIA for France and CWI for the Netherlands[32] was based on the initiative of the research community. This first nucleus of three members quickly opened up. Great Britain in 1990, Portugal and Italy in 1991, Norway, Greece and Sweden in 1992 ... were the first to join a structure that functions essentially as a research network. In 1992 ERCIM became a European Economic Interest Grouping (EEIG). It assumed the coordination of national programs, the management of grant programs and the animation of working groups. If it reveals momentum, the ERCIM's path also highlights all the difficulties of a real European cooperation. Obtaining the status of EEIG was particularly difficult and required long negotiations, which were concluded in the case of Great Britain by a Parliament vote! In 1999, an overview of its activities showed some successes but also the

---

30 Speech by Robert Goebbels at the Eureka Intergovernmental Conference (Hanovre, 5 Novembre 1985), http://www.ena.lu/

31 Idem.

32 *Newsletter* n° 1 *de l'Inria, GMD et du CWI*, avril 1989, archives Inria, boite Ercim.

limits of the effectiveness of a "network" operation, with minimal budgets. "It cannot be denied that with 14 current members and more to come there are problems in achieving consensus on every point, which may well increase as the number increases. The member institutes and consortia can be seen as representative of the IT scenario of their respective country. The diversity of the represented countries and consequently the diversity of opinions means that there is a permanent discussion on the right ongoing strategy for ERCIM."[33] pointed out a newsletter from the institution in 1999. One of the major difficulties was the tension between the more basic research-oriented members of the network and those more oriented towards applied research. "Today, on the one hand ERCIM promotes research and on the other hand it is increasingly involved in the European integration process. It is fascinating to see how some of the problems, but also the benefits of the European Union in the larger context of society are duplicated in the smaller context of research and science.". While underlining the increasingly strong direct impact of ICT on everyday life the redactor finally recalled that: "ERCIM members not only determine how ERCIM will develop in the future, but have their share of responsibility in the process of European integration." [34]. It is understandable that the "science diplomacy" dimension is essential here, whereas innovation diplomacy, which is supposed to be oriented towards industrial systems, cannot really be implemented. By having favored fundamental research too much, at the expense of research better linked to markets and demand, European mechanisms have certainly worked in favor of a fertile European innovation space, but all too often they have allowed third-party interests to take full advantage of it... We can thus see that European start-ups, after the long evolution of projects funded by European Union, have finally been taken over by large American groups...

Despite the limits inherent in the weakness of its resources and its mode of operation, ERCIM is nevertheless an important asset for coherent European action in the field of Information and Communication Technologies. The institution is becoming an element of a European science diplomacy that can, through certain forms of action, be linked to a true innovation diplomacy. Thus, on January 1, 2003, the French

---

[33] ERCIM News, N° 39, October 1999. http://www.ercim.org/publication/Ercim_News/enw39/bierduempel.html

[34] Idem.

Institut National de Recherche en Informatique et en Automatique (Inria) handed over to Ercim the responsibility of being the European host of the World Wide Web Consortium (W3C). This role was thus transferred from a national institution, which had obtained it after real competition with other European research centers in 1995[35], to an organization representing a large part of the researchers of the old continent.

5. Conclusion

Identifying the birth and the evolution of an innovation diplomacy in the field of Information and Communication Technologies, allows us to understand what this concept can concretely cover. It also allows, in a more specific way, to characterize the deep links between the history of this field and the place that Europe currently holds (or does not hold) in the world' s balance of power. Once dominant, European interests were gradually circumscribed by the United States. It was by seizing the opportunities offered by major technological changes that American power was able to act effectively in favor of a redefinition of the terms of the game and to project the action of its best companies on a worldwide scale. This success was long in coming. It was based on the capacity of the American market to produce innovative high-tech companies capable of competing on an international level. The other necessary condition was the establishment of negotiation processes between the different public and private actors to define a common (and often limited) framework for action. Indeed, unlike the field of submarine cables, where the role of the administrations was extremely dominant in a technological framework modified only at the margin over the decades, radio gave a central place to companies in a very unstable technological context, marked by successive and unexpected waves of innovation. The coordination between the actors coming from the public and private spheres became then a crucial point to define a balanced plan and to make a consistent action achieve success.

In the inter-war period, European initiatives appeared to be reasonably effective within the scope of the activities carried out by the British,

---

35 Pascal Griset, and Valérie Schafer. (2011). «Hosting the World Wide Web Consortium for Europe: from CERN to INRIA», *History and Technology*, 27(3), Taylor & Francis, 353–370

French and German interests when they were able to rely on their national diplomacy and, even more so, when they were able to rely on the support of convergent action by their respective national diplomacies. This observation, made for the South American space in the 1920s, is too rarely found afterwards. Later this was true, for example, for the defense of the GSM standard. However, this example is almost an exception. Attempts to build European business alliances, sectoral support "plans", the creation of ecosystems favorable to innovation ... few initiatives have led to tangible results. The weight of the GAFAMs in the present time allows us to note that if the efforts were meritorious, their fruits were very rare. The way the European Union functions and the -sometimes conflicting- visions of the member states further complicates the situation, even though the member states, bound by commitments relating in particular to competition issues within the internal market, can no longer act individually. Consequently, the affirmation of an innovative European diplomacy, which would be capable of defending Europe's strategic and economic interests by supporting the action of European companies working in concert with the authorities of the Union, remains a somewhat distant objective. It is of course possible, as in other areas, to set less demanding short-term goals. Thus, Jos Leijten believes that "One of the important questions will be if innovation diplomacy can play a role in uniting nations around a number of challenges which the world faces, such as climate change, population growth, water shortages, migration, etc."[36]. However well-founded these perspectives may be, they cannot exempt the European Union from the most classic markers of truly effective diplomacy, namely defending the interests, including in the short/medium term, of the community it represents. The field of information and communication technologies underlines the extent to which this objective cannot be decoupled from longer-term objectives. Its history also draws our attention to the indispensable construction of a common base that will allow this diplomatic action to be anchored in a system of references shared by all the actors. Defining the main lines of a European industrial policy, making world-class companies emerge, should be considered as a priority even if it means reconsidering the ultra-competitive rules of a European market that too often abandons dispersed companies to the appetites of globalized groups strongly

---

[36] Leijten, Jos. (2017). "Exploring the Future of Innovation Diplomacy", *European Journal of Futures Research*, 5(20), 1–13.

backed by their own diplomacy. Meeting these challenges cannot be part of a classical economic diplomacy[37]. In a field such as information and communication technologies, it is indeed a true innovation diplomacy that must be evoked. High technologies imply the implementation of specific processes taking into account the complexity of the stakes and the need to have a real expertise that only the actors directly related to the field, scientists, technicians whether they belong to private or public institutions, can mobilize. Exchanges between the different actors involved and the diplomats and more globally civil servants are all the more complex to organize. Its deployment also requires the creation of upstream mechanisms allowing to support the competitiveness of European companies by financing research directly linked to the challenges of international competition. While the paths to be followed to achieve this may be somewhat complex, a strong political will may allow them to appear.

In order to achieve these goals, an effective convergence of the actions carried out by the different DGs of the Commission in close collaboration with the European External Action Service is imperative. If this objective, taken as a whole, seems difficult to achieve, the identification of pilot areas, linked for example to energy or health issues, could constitute a first step. This would allow us to imagine new methods of consultation and action. For this to happen, the politicies will must be strong and no longer be satisfied with announcements. Whether these methods follow the example of Airbus, the European Patent Office or Ariane Espace and are based on intra-European multilateralism, or whether they are part of mechanisms directly supported by the Commission, the determinants of success seem to be historically informed. They include the definition of a clear policy, shared by actors who can converge, through prior negotiation, towards precise objectives. They imply associating private interests with the general interest, without excluding the space for competition and incentives intrinsic to innovation. They require the construction of spaces and procedures for dialogue between public authorities, the private sector and scientists in order to define clear objectives that allow for effective support. The tragic circumstances that Europe is experiencing today can remind us that it is rare, but always advantageous, to take such decisions in a serene manner rather than in the emergency of crises that

---

37 Badel, Laurence. (2018). "Economic Diplomacy", in: Gordon Martel (ed.), *Encyclopedia of Diplomacy*. Hoboken, NJ: Wiley.

unfortunately risk being multiplied in the future. Such an objective can only be achieved by establishing a global and trusting dialogue between European private companies and the EU institutions so that together they can define an ambitious and realistic strategy.

Chapter 4

Entrepreneurial Universities and Science Diplomacy: Harnessing Benefits at Micro, Meso, and Macro Levels in Sweden's Academic Landscape

Magnus Lindqvist and Hans Lundin

Abstract: This report provides a comprehensive analysis of the evolving role of universities in the context of Science, Technology, and Innovation (STI) and their contribution to international affairs, particularly through the lens of the Swedish academic landscape. It delves into the concept of "Entrepreneurial Universities" and their growing influence in the realm of science diplomacy, which extends into research, innovation, and international collaboration. In summary, it encapsulates the strategic roles of universities in STI collaborations and international affairs, highlighting the pivotal roles played by universities in STI collaborations and global affairs, with a deep dive into the Swedish educational framework. It notably showcases the KTH Royal Institute of Technology's strategic approach to internationalization, emphasizing their targeted geographical focus. It underscores the importance of a synergistic approach, combining top-down policy directives with bottom-up academic innovation, facilitated through the meso-level of university administrations, to maximize the impact of science diplomacy. The recommendations aim to enhance global STI engagement, aligning with national innovation, competitiveness, and sustainable development goals.

Keywords: Science Diplomacy, Entrepreneurial University, Multi-level governance, New Diplomacy, International Research Collaboration

Key points:

1. **Interplay between STI and international affairs**: The report underlines the increasing interdependence of the knowledge economy, emphasizing the need for robust international knowledge

networks, scientific collaboration, innovation ecosystems, and value chains.

2. **Critical role of universities in science diplomacy**: Universities are identified not just as educational and research hubs but as pivotal players in innovation and economic development, contributing significantly to science diplomacy through various international collaborations and partnerships.
3. **Balancing top-down and bottom-up approaches**: A major theme of the report is the necessity of balancing a top-down approach in science diplomacy with the bottom-up, research-driven realm of academia. It suggests that effective science diplomacy needs to tap into the grassroots level of research and innovation.
4. **Meso-level as a strategic link**: The report emphasizes the crucial role of the meso-level, found particularly at the university administration level, in bridging the macro (governmental) strategies and the micro (research group and researcher) initiatives. This level acts as a vital connector, ensuring that top-level policies and international STI trends are effectively translated and integrated into the university's operational framework.
5. **Case Study—KTH Royal Institute of Technology**: The report illustrates these concepts through the example of KTH, highlighting its successful internationalization strategies, with a focus on specific geographical regions like Latin America. KTH's approach to managing complexities and enhancing regional understanding is showcased as a model for effective international engagement and development.
6. **Policy recommendations**: The report proposes several recommendations, including fostering multi-level integration, enhancing the role of university administrations, and improving government-university alignment. It highlights the importance of universities using entrepreneurial models to engage in innovation and commercialization, thereby contributing significantly to science diplomacy.

1 Introduction

In the evolving landscape of the knowledge economy, the interplay between Science Technology and Innovation (STI) and International Affairs has become increasingly intricate. Knowledge networks, scientific collaboration, innovation ecosystems, and value chains are deeply intertwined on a global scale. Departing from logics of "mode 2"[1] knowledge production and the "triple helix"[2] policy approach, the role of academic stakeholders are seen as essential actors, alongside industry and other societal actors to foster and maintain national innovation competitiveness, economic growth, and to create sustainable societal solutions.

Consequently, universities role as part of a foreign policy agenda is crucial in many aspects. Not only as a key player to bridge bilateral connections between nations within an agenda of "Science"- or "Innovation Diplomacy"[3], but also as an ingredient to position a nations role in a perceived competitive global environment.

This intricate interconnection highlights the role of academic stakeholders in fostering national innovation competitiveness, economic growth, and sustainable societal solutions. There are several incentives for the individual researcher and universities as institutions to engage in international STI collaboration, but political motives or the "national agenda" is seldom the main motivation driver[4]. Hence, although synergies between governmental support and university-led initiatives can create a dynamic ecosystem fostering sustainable development and technological advancement, policymakers still have to figure out how diplomatic actions, policy instruments, and working methods adapt to motives and rationales for universities to perform and align to overall political agenda and objectives in international collaboration. At the same time, the government has significant potential to capitalize on existing academic and research successes, as well as established networks. By nurturing and reinforcing what already exists, it can lay a stronger foundation for further integration with diplomatic missions and strategic bilateral relationships.

The emergence of the "Entrepreneurial University" model marks a pivotal shift in this dynamic. Universities are not only centres of education and research but also key drivers of innovation and economic development. Their entrepreneurial activities, ranging from research commercialization to industry collaboration, align well with the goals of STI diplomacy, which aims to foster and enhance global knowledge

1 Gibbons et al. (1994).
2 Etzkowitz and Leydesdorff (1995).
3 Leijten (2017).
4 Wagner (2006).

networks. Moreover, the role of universities in STI diplomacy extends beyond traditional academic boundaries. They now act as vital intermediaries and facilitators in the partnership between academia, industry, and government, embodying the "triple helix" model of innovation. This expanded role allows universities to strategically position themselves as key actors in the national and international STI ecosystem, influencing policy, driving innovation, and shaping global STI agendas.

However, this enhanced role creates complexities, especially in decentralized university structures where modes of governance and policies might need to balance the usual top-down approaches of international affairs with the bottom-up characteristics of academia in order to be effective. The majority of international research collaborations stem from informal and interpersonal contacts among researchers, highlighting the need for university administrations to adeptly navigate these dynamic, loosely structured networks[5].

In essence, to be effective and legitimate STI diplomacy, it is crucial to involve academic stakeholders actively. This involvement not only enriches the STI-driven agenda but also ensures that the right expertise and innovative capacities are harnessed for successful international collaboration[6]. In the following section of this policy brief, we explore and illustrate these dynamics within the context of Sweden, a leading innovative and export-driven economy. It delves into the strategic roles universities play in international STI collaborations. With a rather fragmented STI funding landscape structure, including various ministries and Higher Education Institutions (HEI) with strong institutional autonomy the Swedish system presents a unique challenge in maintaining a cohesive discourse on STI diplomacy. The decentralized nature of these institutions suggests that a rigid, top-down approach to science diplomacy is not feasible[7]. Instead, a more inclusive and adaptive approach is required, that acknowledges the different objectives of various stakeholders, and also highlights at which level in the system the different actions and strategies are developed and organized by the actors.

This report argues for the interconnectedness of macro (governmental), meso (institutional management), and micro (research group and

---

5 Wagner and Leydesdorff (2005).

6 Wagner and Graber (2018).

7 STINT (2022).

researcher) levels in the organization of international STI activities and emphasizes the necessity of procedures that acknowledge and integrate these layers. To demonstrate this approach, we examine how KTH implements aspects of its internationalization strategy by concentrating on specific geographic regions. The report particularly focuses on KTH's Latin American initiatives, concluding with policy recommendations that consider this multi-level interconnectedness to enhance global STI engagement.

2. Method and Data Collection

This report outlines a conceptual framework for Science Diplomacy, drawing on existing academic literature in the field. Our empirical analysis is primarily based on data from a survey conducted as part of the Swedish Public Inquiry of Internationalisation "SOU 2018: 3 A Strategic Agenda for Internationalisation," which was distributed to Swedish University Management in 2017. Additionally, we incorporate secondary data, predominantly sourced from the 2022 STINT report titled "Science Diplomacy in and for Sweden".

The authors, practitioners at KTH, leverage their practical experience and expertise to offer a detailed perspective on KTH's operational approaches in Science Diplomacy. While this report does not claim to meet the rigorous standards of academic research, it aims to construct a well-informed argument and offer policy recommendations for the effective administration and implementation of Science Diplomacy. This is particularly focused on the practical workings of universities in the realm of international collaboration and diplomacy in the scientific arena

3. Emerging Paradigms in STI Diplomacy: The Intersection of Academia and Government

A basic concept to grasp the interrelation between the government and academia relates to the evolving concept of "New Diplomacy"[8], where non-state actors like NGOs, international foundations, and multinational industries play a crucial role in a country's diplomatic outreach.

---

8 Hamilton, K., and Langhorne, R. (1995). *The Practice of Diplomacy.*

In a similar way, universities and their researchers emerge as pivotal actors on the global scene. Their global scope and extensive international connections with other STI stakeholders place them at the forefront of a diplomatic agenda that relates to the development or promotion of a nation's STI landscape. The partnership between academia and state actors is both strategic, aligning with national interests and policies, but also bottom-up driven by the decentralized, dynamic nature of academic research and collaboration.

We know from literature that governments play a crucial role in fostering these academic cross-border activities. Policy action primarily consists of allocating specific grants and funds aiming to foster international collaboration, for example through travel grants or bilateral joint-research calls. Also, open up more overarching research grants, such as the European Union's Horizon Europe program, to include international partners. Bilateral and multilateral agreements further facilitate academic exchange and collaboration by creating frameworks for policy dialogue between funding agencies in different countries[9].

Scholarships and research incentives are key in stimulating these activities, making studying abroad and international research projects more feasible. Promotional and branding activities are equally important, attracting international students and academics in a competitive global education market. Another critical aspect is the engagement with the academic and professional diaspora[10]. This not only enriches the domestic academic landscape but also creates networks of international collaborations, essential in the context of STI diplomacy. Adjusting immigration and visa policies is not usually directly linked to STI diplomacy action, but vital to enhance the mobility of international students and researchers and thus promoting cross-border academic activities. Without migration rules to allowing academic cross-border flows, all other related activities and engagement might become redundant.

According to a survey performed by the Swedish Internationalization Inquiry (SOU, 2018, p. 3) most universities have their own internationalisation strategy. For example, all Swedish universities has elaborated some sort of strategic plan to work with internationalisation. These strategies often indicate prioritized countries/regions and define strategic

---

9 Flink and Schreiterer (2010); Boekholt et al. (2009).

10 Larner (2015).

partnerships with international stakeholders such as universities, public institutions, and industry.

The same inquiry shows that while strategic partnerships at universities commonly emerge independently of diplomatic actions, university managements recognize the advantages of aligning with governmental initiatives in STI. Although government-initiated actions such as bilateral STI frameworks and coordinated delegation trips does seldom seem to lead to more concrete strategic STI partnerships for universities, actions like bilateral STI agreements and organized national delegation trips are seen as facilitators for universities' strategic international STI plans. Government involvement not only offers a sense of security and legitimacy for initiating new collaborations, especially in politically or security-sensitive regions, but also elevates the profile of these university endeavours. These initiatives signal to STI stakeholders in target countries that these efforts are of national importance. Furthermore, government actions are instrumental in establishing networks and creating funding opportunities, both domestically and with targeted countries. This is particularly crucial in countries with hierarchical political systems, as it enables access to high-level contacts in new regions.

At the same time, Swedish university management express that government-initiated delegation trips, bilateral STI framework agreements, as well as targeted funding to particular countries can be substantially improved to better align with their work. Swedish university management express that a lack of forum where these things can be discussed and prepared among relevant academic stakeholders and government representatives is missing. Furthermore, university administrators are suggested to be more involved, in order to have action that is more consistent and follow up to international strategic initiatives. Experiences show that leading university management functions do not have the time to implement, neither to focus on long-term commitment to translate strategies into action.

In a separate report[11], STINT (The Swedish Foundation for International Cooperation in Research and Higher Education) analyses and contextualizes the concept of Science Diplomacy within the Swedish STI policy landscape. Likewise, this report advocates for the creation of a more structured forum for STI stakeholder participation. Such a

---

11 STINT (2022).

forum would enable and facilitate meaningful debates and joint initiatives among semi-independent Swedish actors in the realm of Science Diplomacy.

4. Strategic Integration of Academia in the STI Ecosystem: Bridging Sectors and Levels

Today, universities are more than just educational institutions; they are dynamic entities playing a pivotal role in the knowledge economy. Tasked with research, education, and the "third mission" which encompasses innovation and societal engagement, universities have become key drivers in technological advancement, innovation, entrepreneurship, and internationalization. The increasing demands from governments and industries for universities to focus on innovation and the third mission do not diminish the significance of education and research. In fact, these missions are interlinked and more crucial than ever, necessitating close interaction with industry and government.

The way many technical universities strategically work with their surrounding environment can be described within the concept of the "Entrepreneurial University"[12]. The Entrepreneurial University is an institution that can take on several roles in society and in the national innovation system as well as through its rather stable financial advantage when financed by the state, admits its faculty to play an essential role in the successful implementation of triple helix arrangements.

The concept of the "Entrepreneurial University" emphasises how universities actively engaging in innovation and commercialization, which motivates them to foster strong industry partnerships, turning academic research into marketable products and services. Fostering international networks within and outside academia is also prioritised since their curriculum is designed to provide entrepreneurial skills in students, integrating experiential learning and real-world project involvement. Thus, a key feature of Entrepreneurial Universities is their diverse funding sources, including private sector partnerships and commercial activities.

Characterized by their adaptability, these institutions respond swiftly to economic and societal changes, often adopting flexible administrative

12 Etzkowitz (2004).

and academic structures. Thus, with such organization at place, the university has larger capacity to act as an organization actor, bridging and capturing strategic initiatives at the macro level with the needs and expectations at university level[13].

Nevertheless, despite this strategic and organizational approach, universities remain complex entities with a bottom-up driven character, preserving the independence and freedom of the academic faculty. This character of academic freedom ensures that the direction and scope of collaborative outcomes are shaped not primarily by university management, or the university administration, but mainly by individual researchers and research groups. It is therefore essential to be aware that there are still practical limitations of implementing, and potential tensions on how to engage in strategic and policy-led internationalisation activities within the university organization.

Universities engage in international collaborations for numerous reasons. At an institutional (meso-) level, the drive is often towards achieving excellence, prestige, and funding. Internationalization efforts might aim to improve reputation, generate income, develop students and staff, and enhance curricular quality[14]. Although these rationales might not necessarily clash with rationales for individual researchers to engage in academic cross-border activities, research groups and individual researchers operate within a more narrow abstraction level. Their motivations include increasing visibility among peers, sharing costs, raising funds, exchanging ideas, and leveraging shared data, and pay little attention to what university management prescribe, unless not substantial funding or regulation is at hand.

4.1. Managing the Multi-Level Approaches

Except for the mentioned multi-stakeholder (e.g. funding agencies, embassies, academia, industry), multi-issue (education, research, innovation) and multi-disciplinary approach that is necessary in order to build international and dynamic knowledge networks it is also important to mention a fourth dimension, the multi-level setting. The multi-level setting contains of the macro-level (e.g. governments & funding agencies) meso-level (university management/hierarchical/institutional) and

13 Guerrero-Cano (2006).

14 Seeber et al. (2016).

micro-level (individual researchers, teachers and students). All levels are important but have different strengths and purposes. The core functions, and basically all operative actions at a university, takes place at the micro level, while funding sources and overall guidelines and structures are taking place at the macro and meso-level. The vast majority of specific research collaborations activities are performed with a bottom-up characteristic of interpersonal research networks and exchanges. The loosely structured, often changing and dynamic international scene for research networks and exchanges is not easily and effectively left entirely for hierarchal planning and control. Political ties and national prestige do not motivate individual researchers to collaborate, and neither are they often updated on current visions and initiatives occurring on the macro-level.

The meso-level, most commonly through external relations offices, management, and administration can have an important function in bridging the macro level funding sources and act as door openers to the micro-level where the individual researcher at the institution is located. Thus, in this function the meso level can combine great local knowledge of the institution with a vide and broad internal network, with an external monitoring perspective focusing on the macro level environment. In the role as a bridge between these two levels there are several perspectives that needs to be taken into consideration in the operational work of building strong international relations between actors in the STI system, mainly considering: (i) network, (ii) funding, (iii) common rationales.

Focusing on building external **networks**, both within the own country, as well as in foreign settings, identifying and understanding politically driven processes or external stakeholders, not only facilitates access to funding but also opens up valuable opportunities for academic endeavours within the university. Networks is not limited to the external networks that needs to be addressed at the meso-level. The knowledge of the own local and internal network is as important as the external network. Accumulated networks within the meso level function to pool knowledge of potential collaboration partners.

Without **funding** there is not much chance to engage researchers or develop highly intensive research or modalities that are to be used or developed. Depending, on the type of collaboration of course the funding can differ. In Sweden, most projects require freestanding and externa support for wages, travel expenses, and conferences. This accumulative process, mainly, funding can then be connected to concrete and small separate initiatives/projects on the micro-level that can be scaled up and continued in other or even bilateral forums.

Common rationales, identifying and narrowing down the aim and goals of collaboration from the macro-level political ambition and incentive is also important to better understand the different roles and functions that each stakeholder need to achieve. This is particularly important when working across sectors, where organizational structures, motives and approaches to collaboration between academic, industrial and public stakeholders probably differs substantially. In other words, rather abstract goals at the macro level to target particular countries needs to be translated into feasible suggestions of activities, that suits the scope of academia. Similarly, commercial activities and more hands-on expectations from industry to engage in collaborating needs to be redefined to identify "what's in it for the faculty". In other words, in some way the scope for collaboration must be relevant for the students or researchers.

Thus, in the context of STI, universities operate across multiple layers of interaction and influence. These include the macro-level (government), the meso-level (university management and administration), and the micro-level (individual researchers, educators, and students). The university administration has the potential to play a pivotal role as a central node in this structure, effectively bridging these layers to ensure cohesive and productive interactions in STI. Their role in strategic planning, resource allocation, partnership building, and managing the complexities of a decentralized institution is vital in maintaining the dynamism and relevance of universities in the global STI ecosystem. Nevertheless, reaping potential of such framework there is a need for skilled, responsive, and proactive administration teams that can navigate the multifaceted realm of STI, leveraging opportunities, and overcoming challenges for the benefit of their institutions and the broader STI community.

Potential roles and functions of university administration in STI bridging the multiple layers	
1. Role in bridging macro and meso levels	**Policy interpretation and implementation**: University administration can act as an interpreter of macro-level policies and international trends in STI, translating them into actionable strategies at the meso-level. This involves understanding governmental agendas, international STI frameworks, and funding opportunities.

	Strategic planning and alignment: Administrations develop strategic plans that align the university's goals with national and international STI priorities. This requires a nuanced understanding of the broader STI landscape and the university's strengths and opportunities.
	Lobbying and advocacy: University administrations often lobby for policies and resources that benefit their institution and the broader academic community. This involves engaging with policymakers and funding bodies at the macro-level.
2. Connecting the micro levels	**Facilitating research and educational initiatives**: Administrations create and support programs that foster research and education activities aligned with national STI initiatives. With good knowledge about individual researchers and research groups, administrators can also suggest concrete actions to national policymakers to scale up bottom-up interests.
	Resource allocation and support: By effectively managing resources, the administration can ensure that micro-level activities (like individual research projects) are adequately supported, aligning with meso and macro objectives.
	Communication and information dissemination: Administrations play a key role in ensuring that information flows effectively between the micro, meso levels and macro level. This includes disseminating information about funding opportunities, international collaborations, and policy changes.
3. Fostering international and cross-sectoral collaborations	**Building networks and partnerships**: University administrations actively seek and manage partnerships with other academic institutions, industry, and government, fostering a collaborative environment conducive to STI advancement.

	Enhancing visibility and reputation: Through strategic communications and branding, administrations enhance the university's international reputation, attracting collaborations, funding, and talent.
4. **Managing complexities and challenges**	**Navigating decentralization and autonomy**: While universities are often decentralized, with significant autonomy given to individual researchers and departments, the administration needs to balance this with the need for a coordinated approach to STI.
	Adapting to rapid changes in STI: Administrations must be agile and responsive to the fast-paced changes in STI fields, ensuring the institution remains relevant and competitive.

5. Optimizing International STI Strategies: KTH's Regional Approach in Action

As highlighted, academic involvement in international STI activities span across the multiple layers: macro (government), meso (university management), and micro (individual researchers and students). Not only the multi-level complexity, but also the fact that universities faces challenges in coordinating and administering collaborative strategic approaches, which often include education, research, innovation, funding, and administration, which commonly operates within its own silos.

To illustrative an example of how Entrepreneurial Universities can organise administrative structures to effectively bridging these multiple levels and silos, we delve into how KTH Royal Institute of Technology works strategically with a regional focus. As one of Europe's leading technical universities and is known for its contribution to innovation and technology KTH holds significant potential for extensive and diversified contact networks and new collaboration opportunities, particularly in the field of STI. However, as most universities KTH's nature as a bottom-up, decentralized organization poses challenges in coordinating and administering rapidly changing external relations. Similarly, the landscape of international STI collaboration is complex, involving a myriad of actors, further amplifying the decentralized nature of these processes.

To manage this complexity and enhance understanding of different regional and local contexts, KTH has implemented a geographical categorization in its internationalization strategies. Since 2011, KTH has focused on four priority regions for internationalization: Latin America, China, India, and Southeast Asia. These regions were chosen due to its perceived potential in expanding in student recruitment and an identified need to create new networks to reap benefits surging from that these are emerging knowledge societies.

Central coordination of these activities is managed by KTH International Relations Office. Each geographical focus group consists of administrative staff, academic representatives and an assigned budget for activities, which altogether plays a crucial role in maintaining and expanding the university's network in the region. This geographic focus allows KTH to build and maintain networks, understand regional contexts, and capitalize on deeper and more advanced cooperation internationally. Narrowing down the scope geographically also allows to work in an integrated manner with student recruitment, student mobility, international research collaboration and more overarching external relations (e.g. with industry), as opposed to a purely thematic one which is a more traditional way of administering internationalisation activities. In many cases, these activities are highly interrelated. For example, international research collaboration might benefit from exchanging students. Industry networks might not only enable interesting scope and funding for research, but can also generate thesis work for students, or support scholarships for degree seeking students.

Focusing on KTHs strategies in Latin America the goal for KTH is to find ways to institutionalize new cooperation by setting up concrete projects, linking existing initiatives and identify other ways on how to engage KTH students, staff and researchers with actors within academia, industry, public institutions and civil society.

More specifically, the aim of the strategy is to:

- Spread knowledge about KTH in Latin America as a world-leading technical university.
- Maintain and develop strategic collaborations with both academia, business and the surrounding community in the region and to clearly establish these at KTH's schools and institutions.

- Develop closer contacts and projects with businesses and financiers active in the region.
- Increase the number of qualified students from the region within KTH's master's program.
- Increase the number of outgoing students to the region.
- Inform interested research groups and program managers at KTH about opportunities for recruitment and research collaborations in Latin America, many times by targeting funding opportunities and scholarships.

In selecting countries for the region's strategic development plan, factors like exchange potential, research collaboration interest, student recruitment opportunities, growing industry, and alignment with Swedish diplomatic interests were considered. This approach has led to increased collaboration with various actors, including Swedish and foreign embassies, financiers, and public and private research institutes.

In view of the above criteria, most of the activities and budget have been concretized to Brazil, Chile, Colombia and Mexico. Activities can play out quite differently and meet various purposes for KTH as well as providing gains and mutual benefits for both KTH, Latin American partners such as universities and industries.

At the macro level, KTH aligns with Swedish governmental policies and initiatives in STI. This alignment can be expressed through KTH's participation in government-led bilateral STI frameworks and/or delegation trips. Such high-level engagements, while not directly dictating the university's strategy, provide a backdrop against which KTH operates, especially in its international ventures like those in Latin America. KTH's administration, representing the meso-level, plays a crucial role in interpreting these macro-level directives and shaping them into actionable university strategies. The administration's efforts depart from that KTH's activities should resonate with the university's entrepreneurial ethos, as well as the interest of the individual research groups. In other words, government, or funding agency led actions are seen as tools for the university, and not vice-versa.

At the micro-level, a coherent and consistent focus at the meso level fosters an internal network and knowledge based within the university in which individual researchers' interest, or competences, can be aligned with external partners and initiatives proposed in Latin America. In general terms, this include facilitating academic exchanges, research collaborations, and participation in international projects. This has also improved the effectiveness of tapping into activities proposed at the macro level, such as branding events, participation in official delegation trips, or government derived bilateral strategic partnerships.

As the internationalization takes a more strategic approach, the number of actors interested in collaborating with KTH, both domestic and foreign, is also increasing. Swedish embassies, foreign embassies in Sweden, foreign financiers as well as business and public and private research institutes are some examples. This means great potential and opportunity for large and diversified contact networks and new collaboration opportunities for KTH and Swedish educational institutions.

Over the long term, this strategy has proven effective in fostering new and broader international collaborations with various societal actors. KTH's activities in Latin America include educational exchanges, participation in high-level research initiatives, and cooperation with industry and ministries. Some examples of activities are more thought through mobility exchange agreements, double degree initiatives, joint research projects funded by external stakeholders (e.g. by Colombian government, companies, foreign funding agencies and partner universities), extraction of externally funded mobility grants, thesis work opportunities in foreign settings and scholarship schemes provided by foreign stakeholders for international students studying at KTH.

Although KTH engage in actions and activities based on its mission within higher education, research and the "third mission" it has been evident that KTH can contribute to broader Swedish Science Diplomacy efforts, in the following ways:

Branding

Branding events abroad, together with Swedish industry and embassies with the aim of strengthen the brand of Sweden, industries and universities. As the biggest and highest ranked engineering institution in Sweden it is in the interest of KTH to be seen as part of the Swedish Engineering and Innovation context. Of course, this also strengthens the KTH brand in Latin America as a world-leading technical university.

Visibility

Being part of official delegation trips by the Swedish government to countries where KTH holds collaborations can highlight existing projects and collaborations, can be important to scale up, or extend these networks. It can also be of importance for both governmental agencies and KTH to be visible together as stating an example how the university is part of the diplomatic mission or objective in the country of interest, with the potential of finding new partners and funding schemes.

Influence

The Swedish potential to collaborate in a triple helix approach between universities, industries and public actors has been noted by many policy makers in Latin America. Many governments are actively looking for partners to translate their own visions in this regard into action. By showing KTH experiences, active projects, and methodologies for research, education, and innovation, can provide powerful examples for the implementation of such visions.

In this report, KTH Royal Institute of Technology serves as a case study to demonstrate how complex challenges can be effectively managed through innovative structuring of internal processes for a wider scope of internationalization. The purpose here is not to portray KTH as a definitive success story; rather, our focus is on the unique approach KTH employs in its internationalization efforts. It is important to note that our methodological approach does not quantify the degree of success or measure the effectiveness of these strategies.

The implementation of internationalization strategies at KTH has encountered its share of challenges and limitations. A significant concern is the reliance on a small group of key individuals, whose extensive networks the system heavily depends upon. This reliance creates a vulnerability, as the system's effectiveness is tied to these individuals' networking capabilities. Another challenge is effectively harnessing and leveraging a wide range of expertise within the university, given the breadth of activities involved in comprehensive internationalization.

To mitigate these issues, it is crucial to integrate the work focused on specific regions with the efforts of individuals working in more specialized areas of the university administration. This includes teams involved in student recruitment, student mobility, and business liaisons. Additionally, fostering close collaboration with individual researchers and research groups is essential. This ensures that external initiatives are not only aligned with their interests but also integrated seamlessly into their respective administrative structures. Such an approach helps in striking a balance between broader strategic objectives and the specific needs and capabilities within different areas of the university.

6. Policy Recommendations

Based on the arguments laid forward in this report, addressing the interconnectedness of macro (governmental), meso (institutional management), and micro (researcher) levels in the realm of international STI activities. These recommendations aim to optimize the role of universities in a Science Diplomacy landscape, enhancing their contribution to national innovation, competitiveness, and sustainable development, while keeping their core missions intact.

- **Foster multi-level integration:** Prioritize the alignment of actions at the macro (governmental), meso (institutional management), and micro (individual researcher) levels. This integrated approach is crucial for a coherent and effective national strategy in science diplomacy.
- **Improve government-university alignment:** Focus on better aligning government initiatives with university strategies, especially regarding funding, international collaborations, and high-level engagements. This alignment is essential for maximizing the impact of science diplomacy efforts.
- **Enhance university administration's role:** Strengthen the role of university administrations in interpreting and implementing macro-level policies and international STI trends. This would involve strategic planning, lobbying, advocacy, and facilitating research and educational initiatives that align with national and international STI priorities.
- **Utilize Entrepreneurial University models:** Leverage the model of Entrepreneurial Universities to engage actively in innovation and commercialization, fostering strong industry partnerships and turning academic research into marketable products and services.

References

Boekholt, P., Edler, J., Cunningham, P., & Flanagan, K. (2009). *Drivers of international collaboration in research.*

Etzkowitz, H. (2004). The evolution of the entrepreneurial university. *International Journal of Technology and Globalisation,* 1(1), 64–77.

Flink, T. & Schreiterer, U. (2010). Science diplomacy at the intersection of S&T policies and foreign affairs: Toward a typology of national approaches, 37(9), 665–677.

Guerrero-Cano, M., Urbano, D., & Kirby, D. (2006). *A literature review on entrepreneurial universities: An institutional approach.*

Larner, W. (2015). Globalising knowledge networks: Universities, diaspora strategies, and academic intermediaries. *Geoforum,* 59, 197–205.

Leijten, J. (2017). Exploring the future of innovation diplomacy. *European Journal of Futures Research* 5(1), 1–13.

Seeber, M., Cattaneo, M., Huisman, J., & Paleari, S. (2016). Why do higher education institutions internationalize? An investigation of the multilevel determinants of internationalization rationales. *Higher Education* 72, 685–702.

SOU 2018:3. (2018). *A strategic agenda for internationalization.* Swedish public inquiry report series. Stockholm.

STINT. (2022). *Science diplomacy in and for Sweden.* STINT Report Series R 22:01. Stockholm.

Wagner, C. S. & Graber. (2018). *Collaborative era in science.* London: Palgrave Macmillan.

Wagner, C. S. (2006). International collaboration in science and technology: Promises and pitfalls. In *Science and technology policy for development, dialogues at the interface,* (pp. 165–176).

Wagner, C. S. & Leydesdorff, L. (2005). Network structure, self-organization, and the growth of international collaboration in science. *Research Policy,* 34(10), 1608–1618.

Chapter 5

The Protection of Fundamental Rights in International Data Transfers from the Schrems II Case

ALINE BELTRAME DE MOURA[1] AND
GABRIEL HENRIQUE CERON TREVISOL[2]

Abstract: The article focuses on the Schrems II decision by the Court of Justice of the European Union (CJEU) and its implications on international data transfers, particularly between the EU and the US. The Schrems II ruling critically assessed whether data transfer mechanisms like the EU-US Privacy Shield and Standard Contractual Clauses (SCCs) ensure an "essentially equivalent" level of fundamental rights protection as provided within the EU. Following the ruling, which struck down the Privacy Shield for failing to protect EU data subjects from US surveillance, the article investigates the ongoing impact on data controllers and processors, discussing how they must now implement "supplementary measures" to maintain EU-level protections, regardless of data destination. Employing a methodological approach of bibliographical and jurisprudential review, the article aims to delve into the background, critical aspects of the ruling, and discuss broader regulatory challenges and potential solutions for harmonizing global data transfer standards to secure fundamental rights in the digital age.

Keywords: International Data Transfers, Regulatory Harmonization, EU-US Privacy Shield.

---

1 Professor of Private International Law at the Federal University of Santa Catarina (UFSC). Ph.D. in International Law from the Università degli Studi di Milano, Italy. Coordinator of the Center of Research in Private International Law UFSC/CNPq as well as the Jean Monnet Network "Bridge Project", co-funded by the Eramus + Program of the European Commission. Chief-Editor of the Latin American Journal of European Studies.

2 Undergraduate Law Student at the Federal University of Santa Catarina. Researcher PIBIC/CNPq in the field of Law and Policies of the European Union. Member of the Latin-American Center of European Studies (UFSC). Member of the Research Group in Private International Law (NDIPr/UFSC).

1. Introduction

Schrems II is the latest ruling in the series of cases brought to the Court of Justice of the European Union (CJEU) by the Austrian activist Maximilian Schrems against the Irish Data Protection Commissioner (Irish DPC), under the light of the Snowden revelations[3] from the last decade, regarding US surveillance programs.

Judged in July 2020, Schrems II, at large, revolves around the level of protection afforded to the fundamental rights of EU data subjects, in the context of an international data transfer. More specifically, the Court analyzed whether the *Privacy Shield* agreement between the European Union and the United States granted an "essentially equivalent" standard of protection of fundamental rights as compared to the EU, and whether Standard Contractual Clauses (SCCs), in the absence of an adequacy decision, could provide that same standard of protection for EU data subjects.

The Court ultimately struck down the Privacy Shield agreement, understanding that the risk of access, use and retention of EU transferred personal data by US public authorities represented a disproportionate interference to the fundamental rights of EU subjects. On the contrary, the Court understood that the SCC decision was valid. However, controllers and processors, also known as "exporters" and "importers", were required to provide "supplementary measures" to maintain the high level of protection afforded in the EU, independently of where that data was transferred.

Three years post-Schrems II, discussions continue on its enforcement impact for data controllers and processors in transfers. Debates persist on the ruling's consequences for EU-US data transfers, and scholars explore its broader influence on international data transfer regulation. The key question involves the potential for international alternatives to harmonize or standardize global data transfer regulations.

Through a bibliographical and jurisprudential review, this chapter aims to analyze the context behind the Schrems II decision and the most

---

3 Greenwald, G., Macaskill, E., and Poitras, L. (2013). "Edward Snowden: The Whistleblower behind the NSA Surveillance Revelations", *The Guardian*. Available at: https://www.theguardian.com/world/2013/jun/09/edward-snowden-nsa-whistleblower-surveillance.

relevant aspects of the CJEU ruling, including how the aforementioned standard and test of "essential equivalence" of protection of fundamental rights makes the EU an important player in its protection in the digital age; and debate its general impacts and possible international regulatory alternatives to promote interoperability of data transfer rules.

2. Summary of the Facts Surrounding Schrems II

Schrems II represents the latest development in a series of cases initiated by Austrian activist Maximillian Schrems against the Irish Data Protection Commissioner (DPC) under the CJEU. The case, decided in July 2020, is a direct continuation of the first Schrems case, known as Schrems I, ruled in 2015. To understand Schrems II, it's essential to recap the events leading up to it.

In July 2013, Schrems, an Austrian resident and Facebook user, lodged a complaint with the Irish Data Protection Commission (DPC). He invoked the 95/46 Directive, seeking to prohibit Facebook Ireland from transferring his personal data to Facebook Inc. in the United States. Back then, EU residents on Facebook entered into a contract with Facebook Ireland, a subsidiary of Facebook Inc., enabling the transfer of personal data from EU users to servers owned by Facebook Inc. in the U.S. for processing.

Two crucial contextual elements precede the discussion of Schrems I. Firstly, during the time of Schrems I, international data transfers between the European Union and the United States were governed by an agreement established in the early 2000s called Safe Harbour. This agreement permitted uninterrupted data flow between the two parties, endorsed by an adequacy decision from the European Commission, ensuring that the United States upheld an equivalent level of data protection.

The adequacy of the Safe Harbour agreement relied on established data protection principles present in international instruments. The agreement allowed U.S. companies to self-certify compliance with these principles, granting a certain equivalence of rights to European data subjects as stipulated by the 95/46 Directive[4]. However, oversight from

4 Voss, G. (2019). "Obstacles to Transatlantic Harmonization of Data Privacy Law in Context", *Journal of Law, Technology & Policy*, 2, 413, 427–428.

external entities regarding the enforcement of these principles by American companies was minimal.

Responding to the post-9/11 enactment of the "Patriot Act" and related measures that expanded data collection without judicial warrants[5], Maximillian Schrems argued to the Irish Data Protection Commissioner (DPC). He contended that (i) the Safe Harbour agreement did not ensure equivalent rights for EU residents and lacked enforceability on U.S. soil. Additionally, (ii) he argued that U.S. laws and data protection practices did not provide sufficient protection for EU personal data stored in the USA, especially considering surveillance activities by U.S. public entities revealed by Edward Snowden. Essentially, Schrems questioned the adequacy status granted by the European Commission to the American legal system through the Safe Harbour agreement (Adequacy Decision 2000/520).

Initially rejected by the DPC, who claimed it lacked the authority to question the adequacy status of a country deemed adequate by the European Commission, Schrems appealed to the Irish High Court, challenging the rejection of his complaint.

In response to a case brought by Maximillian Schrems challenging the Safe Harbour agreement, the Irish High Court referred the matter to the Court of Justice of the European Union (CJEU) in July 2014. In October 2015, the CJEU invalidated the adequacy decision (2000/520) underlying Safe Harbour.

In Schrems I, the CJEU's decision centered on two key aspects. Firstly, it acknowledged the potential compatibility of a self-certification system with EU data protection standards but stressed the importance of effective mechanisms for detection and supervision. The court observed that U.S. public authorities were not obligated to follow Safe Harbour principles, and the European Commission's adequacy decision lacked clarity on how the U.S. ensured an "adequate" level of protection[6]. The CJEU emphasized that adequacy decisions should adhere to a higher standard of protection, guided by the European Charter of Fundamental Rights (ECFR) and the concept of "essential equivalence" to EU standards.

---

5 Voss, G. (2019). "Obstacles to Transatlantic Harmonization of Data Privacy Law in Context", *Journal of Law, Technology & Policy*, 2, 428.

6 Case C-362/14, Schrems I, para. 83.

The Court highlighted that the data protection principles in the Safe Harbour agreement could be set aside for reasons of national security, public interest, or law enforcement by U.S. authorities. The agreement explicitly stated that if U.S. law conflicted with its principles, U.S. organizations, whether part of Safe Harbour or not, must comply with American law, giving precedence to U.S. law[7] over CFR in such conflicts. Additionally, there was no assurance in the agreement or the adequacy decision that U.S. laws limited the power of public authorities to interfere with the fundamental rights of EU data subjects[8].

Subsequently, the High Court annulled the decision rejecting Schrems' complaint and referred the case back to the Irish Data Protection Commissioner (DPC)[9]. During a new inquiry, Facebook Ireland revealed that, after the Safe Harbour invalidation, most EU-collected personal data was transferred to Facebook Inc. using standard contractual clauses (SCCs) introduced by the European Commission in 2010[10]. With these new details, the Commissioner asked Schrems to revise his initial complaint.

In December 2015, Schrems filed a new complaint, asserting that U.S. law still obligated Facebook Inc. to provide transferred personal data to U.S. authorities, violating EU fundamental rights. He argued that the use of transferred EU personal data in state surveillance activities violated specific articles of the European Charter on Fundamental Rights. Schrems also contended that the standard contractual clauses did not provide adequate protection for EU personal data. Once again, he requested the DPC to prohibit or suspend the transfer of his personal data to Facebook Inc.

In May 2016, the Irish Data Protection Commissioner (DPC) agreed with Schrems' allegations that personal data from EU subjects transferred to the U.S. was at "serious risk" of being accessed and processed

---

7 Case C-362/14, Schrems I, para. 84 and 86.

8 Case C-362/14, Schrems I, para. 88.

9 Kuner, C. (2020). "International data transfers, standard contractual clauses, and the Privacy Shield: the AG opinion in Schrems II" *European Law Blog*. Available at: https://europeanlawblog.eu/2020/01/07/international-data-transfers-standard-contractual-clauses-and-the-privacy-shield-the-ag-opinion-in-schrems-ii/.

10 Decision n. 2010/87: Commission Decision of 5 February 2010 on standard contractual clauses for the transfer of personal data to processors established in third countries under Directive 95/46/EC.

by U.S. authorities, contrary to EU fundamental rights. It found that the 2010 Standard Contractual Clauses (SCCs) could not address this risk, as they only provided contractual and procedural remedies against EU exporters and third-country importers. EU data subjects could not challenge U.S. surveillance programs in American courts.

In July 2016, the European Commission introduced Privacy Shield as the successor to the Safe Harbour agreement, maintaining the adequacy status for the U.S. Despite some improvements, critics called it "*Safe Harbour* 0.7" due to unresolved issues, especially regarding U.S. surveillance programs[11]. Schrems updated his complaint to challenge the adequacy of Privacy Shield. In May 2018, the Irish High Court, at the request of the Irish DPC, referred eleven preliminary questions to the CJEU.

These issues can be summarized in five major groups of subjects[12]: (i) whether the GDPR applied to data transfers by an economic operator in the Union to another established in a third country, when the data was likely to be processed in a third country for security and law enforcement purposes; (ii) what was the level of protection applied to the data transferred through SCCs, and whether that level of protection was to be read in the light of the ECFR; (iii) whether the SCCs were valid in the light of articles 7, 8 and 47 of the ECFR; (iv) what was the role of DPAs when supervising the application of SCCs, and (v) whether the *Privacy Shield* ensured an adequate level of protection, according to the GDPR, read in the light of the EU Charter of Fundamental Rights.

3. Important Aspects of the Ruling

All the five groups of subjects mentioned above will be summarized in this session. In addressing the first group of subjects, the Court reiterated the interpretation from Schrems I that international data transfers are encompassed within the definition of processing according to Article 4(2) of the GDPR[13]. As a result, all data transfers were considered within the material scope of the regulation under Article 2(1).

---

11 Rotenberg, M. (2020). "Schrems II, from Snowden to China: Toward a New Alignment on Transatlantic Data Protection", *European Law Journal*, 26(1–2), 5.

12 Costello, R. A. (2020). "Schrems II: Everything Is Illuminated?", *European Papers Forum*, 5(2), 4.

13 Case C-311/18 Schrems II, para. 82–83.

The Court stated that the only potential exception would be if the transfer between Facebook Ireland and Facebook Inc. fell under one of the exceptions[14] to processing outlined in Article 2(2) of the GDPR. However, the Court emphasized that these exceptions primarily pertained to activities in which individuals are involved, such as purely personal or household activities, which did not apply to the mentioned transfer.

Furthermore, the Court highlighted that, based on the wording of Article 45(2)(a) of the GDPR, processing of personal data by a third country for public security, defense, and state security purposes could not exclude such a transfer from the application of the GDPR[15]. This is because such legislation is considered by the European Commission when assessing the level of protection provided by that third country.

Therefore, since the transfer between economic operators did not fall within any exceptions to processing under the regulation, and legislation related to public security, defense, and national security needed to be considered by the European Commission for adequacy decisions, the Court concluded that the transfer fell within the scope of the GDPR[16].

Concerning the second and third groups, the Court[17] emphasized that Article 46(1) of the GDPR, which refers to "appropriate safeguards," "enforceable rights," and "effective legal remedies," should be interpreted in the context of Article 44 of the GDPR. Article 44 specifies that the level of protection provided by the regulation should not be undermined during the transfer, regardless of the legal basis used by the data exporter.

The Court also clarified[18] that, following Schrems I, a third country must ensure, based on its domestic law or international commitments, an "essentially equivalent" level of protection for fundamental rights and freedoms as guaranteed within the EU by the GDPR, read in the light of the EU Charter of Fundamental Rights. Even in the absence of an adequacy decision, controllers and processors, when using Standard Contractual Clauses (SCCs), should aim to achieve the same level of protection[19] for data transferred, regardless of the destination.

---

14 Case C-311/18 Schrems II, para. 85.
15 Case C-311/18 Schrems II, para. 87.
16 Case C-311/18 Schrems II, para. 89.
17 Case C-311/18 Schrems II, para. 92.
18 Case C-311/18 Schrems II, para. 94.
19 Case C-311/18 Schrems II, para. 95–96, 101.

The Court delegated the assessment of the standard of essential equivalence to the entities involved. This assessment should consider both the contractual clauses agreed upon and the relevant aspects of the legal system in the third country[20], especially those outlined in Article 45(2) of the GDPR.

Additionally, SCCs are designed to be applicable in the absence of an adequacy decision. In such cases, it is the responsibility of the controller or processor in the EU to take measures to compensate for the lack of data protection in the third country, ensuring compliance with data protection requirements and the rights of data subjects within the EU, including the availability of enforceable data subject rights and effective legal remedies[21].

The Court highlighted that Standard Contractual Clauses (SCCs) alone, due to their contractual nature[22] binding only the parties directly involved, may not always be sufficient to guarantee compliance with the required level of protection in the EU[23]. To address this limitation, the Court emphasized that controllers and processors may need to implement additional safeguards, referred to as "supplementary measures," to ensure the "essential equivalence" standard of protection is maintained.

The validity of the SCC decision from 2010, according to the Court, depends on whether it incorporates effective mechanisms that make the "essentially equivalent" standard possible[24]. The Court concluded that the SCC decision remains valid, but controllers and processors must implement "supplementary measures" to ensure a high standard of protection. The ruling did not specify the nature of these measures.

Subsequently, in November 2020, the European Data Protection Board released recommendations on supplementary measures,

---

20 Case C-311/18 Schrems II, para. 104–105.

21 Recital 108 and 114, of the Regulation (EU) 2016/679 (General Data Protection Regulation—GDPR).

22 Costello, R. A. (2020). "Schrems II: Everything Is Illuminated?", *European Papers Forum*, 5(2), 10.

23 Drechsler, L. and. Kamara, I. (2022). "Essential Equivalence as a Benchmark for International Data Transfers After Schrems II", in: Kosta, E. and Leenes, R. (eds.), *Research Handbook on EU Data Protection*. Cheltenham: Edward Elgar Publishing, pp. 25–26.

24 Case C-311/18 Schrems II, para. 137.

categorizing them as contractual, organizational, and technical[25]. The emphasis was on combining different measures, especially technical ones like encryption, pseudonymization, and multi-party computation[26], to enhance the level of data protection during transfers[27].

Reiterating the stance from Schrems I, the Court maintained that national Data Protection Authorities (DPAs) must suspend or prohibit international data transfers if contractual clauses cannot be fulfilled or if the required level of protection cannot be guaranteed in any other way. DPAs, since Schrems I, play a crucial role in supporting international data transfers on a contractual basis, as the standard contractual arrangements only provide a legal basis if national authorities ensure compliance with EU law[28].

In the context of the fifth group, the Court's analysis of Privacy Shield's adequacy can be divided into several aspects. Firstly, the Court observed that the Privacy Shield allowed limitations on the application of data protection principles in the United States in the interest of national security, public interest, or law enforcement. This primacy of American law meant that self-certified American organizations receiving EU personal data had to deviate from the agreement's principles when they conflicted with American law and proved incompatible.

According to the Court, such derogations could lead to access and use of transferred personal data by American public authorities, particularly within surveillance programs like PRISM[29] and UPSTREAM[30],

---

25 EUROPEAN DATA PROTECTION BOARD. Recommendations 01/2020 on measures that supplement transfer tools to ensure compliance with the EU level of protection of personal data, para. 98, 128 and 77.

26 Drechsler, L. and. Kamara, I. (2022). "Essential Equivalence as a Benchmark for International Data Transfers After Schrems II", in: Kosta, E. and Leenes, R. (eds.), *Research Handbook on EU Data Protection*. Cheltenham: Edward Elgar Publishing, p. 26.

27 European Data Protection Board. Recommendations 01/2020 on measures that supplement transfer tools to ensure compliance with the EU level of protection of personal data, para. 47.

28 Rotenberg, M. (2020). "Schrems II, from Snowden to China: Toward a new alignment on transatlantic data protection", *European Law Journal*, 26(1–2), 8.

29 Costello, R. A. (2020). "Schrems II: Everything Is Illuminated?", *European Papers Forum*, 5(2), 7.

30 Case C-311/18, Schrems II, 62.

based on Section 702[31] of FISA and Executive Order 12.333 on 1981. This created the potential for interferences with the fundamental rights of individuals whose data was transferred from the EU to the USA.

However, the Court clarified that, at first glance, interferences with the right to respect for private life (Article 7 of the ECFR) and the right to data protection (Article 8 of the ECFR) were not inherently inadmissible[32]. As established in Schrems I, international data transfers were considered a form of processing, constituting an interference with the fundamental rights of the Union. Nevertheless, certain interferences were allowed as long as they passed a proportionality test outlined in Article 52(1) of the ECFR.

The proportionality test became the measure of the essential equivalence standard applicable to all GDPR transfer mechanisms. To maintain this standard during transfers, all elements surrounding the transfer, from the legal situation of the third country to the technological measures used, should be considered.

The literature[33] explains that the Court has established four main criteria to analyze the proportionality of the transfer:

(i) Any interference caused by transfers has to be provided by a clear and precise law, and its application should be foreseeable to the people subjected to it;
(ii) The transfer needs to be justifiable by an objective of general interest recognized by the EU or by the necessity to protect rights and freedoms of others;
(iii) The transfer must respect the essence[34] of fundamental rights;
(iv) The transfer must be the adequate measure to meet its goal and it must be the one that less affects the involved rights (necessity and proportionality in the strict sense).

---

31 Section 702 Overview. Available at: https://www.dni.gov/files/icotr/Section702-Basics-Infographic.pdf.

32 Costello, R. A. (2020). "Schrems II: Everything is Illuminated?", *European Papers Forum*, 5(2), 6–7.

33 Drechsler, L. and. Kamara, I. (2022). "Essential Equivalence as a Benchmark for International Data Transfers After Schrems II", in: Kosta, E. and Leenes, R. (eds.), *Research Handbook on EU Data Protection*. Cheltenham: Edward Elgar Publishing, pp. 14–20.

34 Lenaerts, K. (2019). "Limits on Limitations: The Essence of Fundamental Rights in the EU", *German Law Journal*, 20(6), 781.

The CJEU, in cases such as Schrems I and Opinion 1/15[35], applied criteria to assess and compare foreign laws with EU standards outlined in the GDPR and the ECFR. This analysis identified three[36] distinct interferences with the fundamental rights of the EU in the context of international data transfers, specifically under the Safe Harbour and the EU-Canada PNR agreement.

First, the right to respect for private life (Article 7 of the ECFR) faced interference when communication data, both metadata and content data, were accessed and retained by the public authorities in the destination country. The Court distinguished between interference in data access and interference in excessive data retention.

Second, the right to data protection (Article 8 of the ECFR) experienced interference due to the transfer itself, as the operation was considered a form of data processing. Lastly, the right to non-discrimination (Article 21 of the ECFR) encountered interference when certain special categories of data were transferred and stored without distinction.

In Schrems II, the CJEU employed a similar line of reasoning and applied a test to assess whether access and retention of EU data by US public authorities constituted interference with certain fundamental rights of the European Union (Articles 7, 8, and 47 of the ECFR). This test played a crucial role in determining the validity of the Privacy Shield agreement[37].

The Court concluded that interferences resulting from surveillance programs based on Section 702 of FISA and Executive Order 12.333 did not meet the requirements for the USA to ensure, subject to the principle of proportionality, a level of protection essentially equivalent to that guaranteed by the EU through its Charter of Fundamental Rights[38].

---

35 European Union. Opinion 1/15—–PNR Agreement UE-Canada, Opinion of the Court (Grand Chamber) of 26 July 2017.

36 Drechsler, L. and. Kamara, I. (2022). "Essential Equivalence as a Benchmark for International Data Transfers After Schrems II", in: Kosta, E. and Leenes, R. (eds.), *Research Handbook on EU Data Protection*. Cheltenham: Edward Elgar Publishing, p. 15.

37 Drechsler, L. and. Kamara, I. (2022). "Essential Equivalence as a Benchmark for International Data Transfers After Schrems II", in: Kosta, E. and Leenes, R. (eds.), *Research Handbook on EU Data Protection*. Cheltenham: Edward Elgar Publishing, pp. 20–21.

38 Case C-311/18, Schrems II, para. 178.

Regarding surveillance programs based on Section 702 of FISA, the Court noted that the Foreign Intelligence Surveillance Court (FISC), responsible for judicial oversight, did not authorize individual surveillance measures[39] but approved broad programs (like PRISM and UPSTREAM) based on annual certifications. The FISC focused on verifying if the programs aligned with the purpose of obtaining external information, without checking whether individuals were appropriately targeted for that purpose[40].

Section 702 of FISA lacked disclosure of limitations to surveillance programs for obtaining external information and did not guarantee clear and precise rules, violating the principle of proportionality. Concerning surveillance programs under Executive Order 12333, the Court noted the absence of opposable rights for foreigners against US authorities in US courts[41]. While acknowledging the existence of Presidential Policy Directive 28 (PPD-28) imposing limitations, announced by the President Barack Obama just a few days after the Snowden revelations, it highlighted that the directive still allowed large-scale data collection without specific targets[42], lacking enforceable rights for EU data subjects against US authorities. This directive was intended to impose limitations on the "signals intelligence activities[43]", linked to the American intelligence services. The Court concluded that neither Section 702 nor Executive Order 12333, with PPD-28, met minimum requirements of proportionality, as the surveillance programs were not strictly necessary[44].

The CJEU decided that neither Section 702 of FISA nor Executive Order 12333, read with PPD-28, met the minimum requirements of proportionality, as the surveillance programs were not strictly necessary[45]. After the proportionality analysis regarding the interference of

39 Foreign Intelligence Surveillance Court (FISC). Available at: https://epic.org/privacy/surveillance/fisa/fisc/#:~:text=The%20Foreign%20Intelligence%20Surveillance%20Court,a ctivities%20in%20a%20classified%20setting.

40 Case C-311/18 Schrems II, para. 179.

41 Case C-311/18 Schrems II, para. 182.

42 Case C-311/18 Schrems II, para. 183.

43 Presidential Policy Directive 28/14. Available at: https://www.fbi.gov/file-repository/ppd-28-policies-procedures-signed.pdf.

44 Case C-311/18 Schrems II, para. 184.

45 Case C-311/18 Schrems II, para. 184.

US surveillance programs on article 7 and 8 of the ECFR, the Court proceeded to analyze whether the *Privacy Shield*, through its Ombudsman[46], guaranteed the fundamental right to an effective remedy before an independent and impartial court, in accordance with Article 45(2)(a) of the GDPR and Article 47 of the ECFR.

Fort the Court, the answer was negative. The Ombudsman mechanism was not sufficient to constitute an effective safeguard for the purposes of the Article 47 of the ECFR, because (i) it could provide such remedies based on actionable rights; (ii) it was unable to issue binding decisions aimed at US intelligence services, and (iii) it wasn't fully independent, insofar as it was appointed by the Secretary of State and was a part of the US State Department[47].

Taking all the elements described above into account, the CJEU invalidated the Privacy Shield[48]. In summary, the decision was based on four factors[49]: (i) the primacy of US criminal law enforcement requirements over the data protection principles of the Privacy Shield; (ii) the lack of necessary limitations and safeguards against the power of US public authorities, particularly in the light of the proportionality test; (iii) the lack of corrective legal measures, both judicial and administrative, to be used by EU data subjects in the US, and, finally, (iv) the deficiencies of the Ombudsman mechanism for the agreement. All these elements prevented the agreement from providing an essentially equivalent standard of fundamental rights for the EU data subjects.

4. Schrems II, Fundamental Rights, and International Data Protection Law

The Schrems II decision is a landmark ruling in EU data protection, particularly concerning the unparalleled substantial EU-US data transfer relationship[50]. This relationship, marked by the dominance of US

---

46 Case C-311/18 Schrems II, para. 194.

47 Case C-311/18 Schrems II, para. 195.

48 Case C-311/18 Schrems II, at para. 199–201.

49 Kuner, C., Bygrave, L. and Docksey, C. (2021). *The EU General Data Protection Regulation (GDPR)*: A Commentary. Oxford: Oxford University Press, p. 784.

50 Murphy, M. H. (2021) "Assessing the Implications of Schrems II for EU-US Data Flow", *International & Comparative Law Quarterly*, 71.

big tech companies and a significant EU consumer market, highlights the decision's importance (that amounts to, approximately, US$ 900 billion[51] in digital cross-border commerce).

The ruling underscores the CJEU's consistent support for a robust data protection framework rooted in fundamental rights[52]. Schrems II emphasizes the non-negotiability of privacy and data protection rights, especially in the context of data transfers[53]. Any lawful restrictions must be specific, strict, and proportionate.

The invalidation of a second data transfer agreement raised concerns about the future of EU-US data transfers, but a new agreement announced in early 2022 addresses these issues[54]. The framework introduces changes[55], including safeguards for surveillance activities, a two-tier redress mechanism with an independent US Data Protection Review Court, and administrative procedures for overseeing US intelligence activities. These adjustments aim to alleviate concerns raised by Schrems II and provide a foundation for ongoing EU-US data transfers.

Scholars are divided on the global impact of the Schrems II ruling and potential solutions for easing cross-border data flows. The lack of a comprehensive international framework for data transfers leads to reliance on domestic laws, regional conventions, or free trade agreements[56]. Despite this, the EU, particularly through the GDPR, is seen as a global

---

51 EUROPEAN COMMISSION. Trans-Atlantic Data Privacy Framework—–Factsheet, 2022. Available at: https://ec.europa.eu/commission/presscorner/detail/en/fs_22_2100.

52 Christakis, T. (2020). "After Schrems II: Uncertainties on the Legal Basis for Data Transfers and Constitutional Implications for Europe", *European Law Blog*, 21.

53 Pierucci, A. and Walter, J. P. (2020). "Better Protecting Individuals in the Context of International Data Flows: The Need for Democratic and Effective Oversight of Intelligence Services—–Joint Statement", *Council of Europe*. Available at: https://rm.coe.int/statement-schrems-ii-final-002-/16809f79cb#:~:text=In%20its%20judgment%20%E2%80%9CSchrems%20II,the%20Charter%20of%20Fundamental%20Rights.

54 European Union. Trans-Atlantic Data Privacy Framework. Factsheet, March 2022. Available at: https://ec.europa.eu/commission/presscorner/detail/en/fs_22_2100.

55 Pesmazoglou, A., and Ahremar, K. (2022). "Schrems II Resolved? Unpacking the EU-US Trans-Atlantic Data Privacy Framework", *International Network of Privacy Law Professionals*. Available at: https://inplp.com/latest-news/article/schrems-ii-resolved-unpacking-the-eu-us-trans-atlantic-data-privacy-framework/.

56 Greenleaf, G. (2021). "Global Data Privacy Laws 2021: Uncertain Paths for International Standards", *Privacy Laws & Business International Report*, (169), 6–7.

influencer and an informal "gold standard"[57] in data protection. Proposed solutions fall into two groups: those promoting international regulatory interoperability with a broader scope of action.

Some scholars suggest the possibility of an international multilateral legally binding instrument to standardize rules for data transfers, but the feasibility and immediate conclusion of such a convention are uncertain[58]. In the absence of this, an alternative proposed in the literature[59] is to strengthen international regional binding instruments, such as the Council of Europe's Convention 108+[60] and the African Union's Convention on Cyber Security and Personal Data Protection. Additionally, the G7 introduced the "Data Free Flow with Trust"[61] initiative after the ruling, aiming to facilitate data transfers between trusted parties adhering to common standards[62]. This initiative seeks a political consensus not only on security and privacy rules[63] but also on a broader agenda of trade policy, regulatory cooperation, and business practices to enable cross-border data flow while accommodating domestic policy preferences and objectives[64].

---

57 Greenleaf, G. (2021). "Global Data Privacy Laws 2021: Uncertain Paths for International Standards", *Privacy Laws & Business International Report*, (169), 4–5.

58 Kuner, C. (2009). "An International Legal Framework for Data Protection: Issues and Prospects" *Computer Law & Security Review*, 25.

59 Robinson, L., Kizawa, K. and Ronchi, E. (2021). "Interoperability of Privacy and Data Protection Frameworks", *OECD Going Digital Toolkit Note*, 21, Available at: http://goingdigital.oecd.org/data/notes/No21_ToolkitNote_PrivacyDataInteroperability.pdf.

60 The Convention 108+ was adopted in 2013 and is considered a modernized version of the Convention 108, first adopted in 1980.

61 Kudo, F., Sakaki, R., and Soble, J. (2022). "Every Country Has Its Own Digital Laws. How Can We Get Data Flowing Freely Between Them?", *World Economic Forum*. Available at: https://www.weforum.org/agenda/2022/05/cross-border-data-regulation-dfft/.

62 Greenleaf, G. (2022). "Default Whitelists: Networking Trust to Reduce Personal Data Export Problems [PPT Presentation], Panel: Regulation of Global Data Flows: A Story of the Impossible", *Computers Privacy and Data Protection (CPDP) Conference*, p. 3.

63 Kudo, F., Sakaki, R., and Soble, J. (2022). "Every Country Has Its Own Digital Laws. How Can We Get Data Flowing Freely Between Them?", *World Economic Forum*. Available at: https://www.weforum.org/agenda/2022/05/cross-border-data-regulation-dfft/.

64 Doherty, S. and Samans, R. (2020). "Foreword in Data Free Flow with Trust (DFFT): Paths towards Free and Trusted Data Flows", *World Economic Forum*, p. 4.

At last, the use of default whitelists as a unified mechanism for data transfers has also been discussed. For this proposal, the best alternative would be for countries to enact whitelists containing other countries that have adopted high standards of data protection, either through accession to the Convention 108+ or through an EU adequacy decision. Or even, for a third-party organization to enact such whitelists. For the literature[65], this method would streamline the verification of high data protection standards and ease data transfers. In the same light, these lists would create a network effect, increasing the value of guaranteeing a high standard of data protection.

On the other hand, the second group is related to initiatives that involve the private sector, that aim to promote interoperability in a more focused scale. Beyond the already mentioned SCCs (model contractual clauses), the GDPR offers other similar alternatives such as (i) binding corporate rules, that are uniformly applicable to all entities of a certain business group; (ii) codes of conduct[66], in order to help categories of controllers or processors[67] apply general data protection provisions, and (iii) certification mechanisms[68], that demonstrate the existence of appropriate safeguards provided by controllers or processors outside the EU in a single transfer or in a set of transfers.

In addition, outside the EU, one example is the Cross-Border Privacy Rules (CBPR) System, developed by all 21 countries that compose the Asia-Pacific Economic Cooperation (APEC). As explained in the literature, under this system, companies are certified for their compliance

---

65 Greenleaf, G. (2022). "Default Whitelists: Networking Trust to Reduce Personal Data Export Problems [PPT Presentation], Panel: Regulation of global data flows: a story of the impossible", *Computers Privacy and Data Protection (CPDP) Conference*, p. 3

66 Robinson, L., Kizawa, K., and Ronchi, E. (2021). "Interoperability of privacy and data protection frameworks", *OECD Going Digital Toolkit Note*, 21, Available at: http://goingdigital.oecd.org/data/notes/No21_ToolkitNote_PrivacyDataInteroperability.pdf, p. 19, 21–22.

67 European Data Protection Board. Guidelines 04/2021 on Codes of Conduct as tools for transfers, 2022. Available: https://edpb.europa.eu/system/files/2022-03/edpb_guidelines_codes_conduct_transfers_after_public_consultation_en_1.pdf, p. 6.

68 European Data Protection Board. Guidelines 04/2021 on Codes of Conduct as tools for transfers, 2022. Available at: https://edpb.europa.eu/system/files/2022-03/edpb_guidelines_codes_conduct_transfers_after_public_consultation_en_1.pdf, p. 9.

with the APEC Privacy Framework as it relates to requirements for the protection of personal data. And, once certified, the companies can transfer data to other CBPR-certified companies[69].

The general effects of the Schrems II decision are still in development. Nonetheless, it is possible to assert that they surpass the territory of the EU. By expanding the essential equivalence standard of protection to all transfer mechanisms contained in the GDPR, more specifically to SCCs, the CJEU directly impacted the data transfer relationships between entities established in third countries with others established in the EU. In essence, this change forces all entities that want to continue trading with the EU to adapt to the requirements provided by the Union, for the better (increasing the level of protection afforded to data subjects that have direct contact with that entity) or for worse (increased spending on cybersecurity and legal services).

As pointed out in the literature[70], the GPDR and its transfer rules have become "a vehicle to assert EU fundamental rights in an international context", directly influencing the state of international data protection law. Now, it can be said that the CJEU rulings, such as Schrems II, have occupied the same level of importance, since by interpreting EU law, the Court can now directly influence matters related to public policy and international relations.

References

Christakis, T. (2020). After Schrems II: Uncertainties on the legal basis for data transfers and constitutional implications for Europe. *European Law Blog*, 21.

Costello, R. A. (2020). Schrems II: Everything is illuminated?. *European Papers Forum*, 5(2).

---

69 Robinson, L., Kizawa, K., and Ronchi, E. (2021). "Interoperability of Privacy and Data Protection Frameworks", *ECD Going Digital Toolkit Note*, 21, Available at: http://goingdigital.oecd.org/data/notes/No21_ToolkitNote_PrivacyDataInteroperability.pdf, pp. 19–20.

70 Kuner, C. (2018). "International Agreements, Data Protection, and EU Fundamental Rights on the International Stage: Opinion 1/15 (EU-Canada PNR) of the Court of Justice of the EU", *Common Market Law Review*, 55(3), n. 3.

Doherty, S. & Samans, R. (2020). Foreword in Data Free Flow with Trust (DFFT): Paths towards free and trusted data flows. *World Economic Forum.*

Drechsler, L. &. Kamara, I. (2022). Essential equivalence as a benchmark for international data transfers after Schrems II. In E. Kosta & R. Leenes (eds.), *Research handbook on EU data protection.* Cheltenham: Edward Elgar Publishing.

Greenwald, G., Macaskill, E., & Poitras, L. (2013). Edward Snowden: The whistleblower behind the NSA surveillance revelations. *The Guardian.* Available at: https://www.theguardian.com/world/2013/jun/09/edward-snowden-nsa-whistleblower-surveillance.

Greenleaf, G. (2022). Default Whitelists: Networking trust to reduce personal data export problems [PPT Presentation], Panel: Regulation of global data flows: A story of the impossible. *Computers Privacy and Data Protection (CPDP) Conference.*

Greenleaf, G. (2021). Global Data Privacy Laws 2021: Uncertain paths for international standards. *Privacy Laws & Business International Report*, (169).

Kuner, C., Bygrave, L., & Docksey, C. (2021). *The EU General Data Protection Regulation (GDPR)*: A commentary. Oxford: Oxford University Press.

Kuner, C. (2020). International data transfers, standard contractual clauses, and the privacy shield: The AG opinion in Schrems II. *European Law Blog.* Available at: https://europeanlawblog.eu/2020/01/07/international-data-transfers-standard-contractual-clauses-and-the-privacy-shield-the-ag-opinion-in-schrems-ii/.

Kuner, C. (2018). International agreements, data protection, and EU fundamental rights on the international stage: Opinion 1/15 (EU-Canada PNR) of the court of justice of the EU. *Common Market Law Review*, 55(3).

Kuner, C. (2009). An international legal framework for data protection: Issues and prospects. *Computer Law & Security Review*, 25.

Lenaerts, K. (2019). Limits on limitations: The essence of fundamental rights in the EU. *German Law Journal*, 20(6).

Murphy, M. H. (2021). Assessing the implications of Schrems II for EU-US data flow. *International & Comparative Law Quarterly*, 71.

Pierucci, A. & Walter, J. P. (2020). Better protecting individuals in the context of international data flows: The need for democratic and effective oversight of intelligence services—Joint statement. *Council of Europe.*

Available at: https://rm.coe.int/statement-schrems-ii-final-002-/16809f79cb#:~:text=In%20its%20judgment%20%E2%80%9CSchrems%20II,the%20Charter%20of%20Fundamental%20Rights.

Rotenberg, M. (2020). Schrems II, from Snowden to China: Toward a new alignment on transatlantic data protection. *European Law Journal*, 26 (1–2).

Voss, G. (2019). Obstacles to transatlantic harmonization of data privacy law in context. *Journal of Law, Technology & Policy*, 2.

Kudo, F., Sakaki, R., & Soble, J. (2022). Every country has its own digital laws. How can we get data flowing freely between them? *World Economic Forum*. Available at: https://www.weforum.org/agenda/2022/05/cross-border-data-regulation-dfft/.

Pesmazoglou, A., & Ahremar, K. (2022). Schrems II resolved? unpacking the EU-US trans-Atlantic data privacy framework. *International Network of Privacy Law Professionals*. Available at: https://inplp.com/latest-news/article/schrems-ii-resolved-unpacking-the-eu-us-trans-atlantic-data-privacy-framework/.

Robinson, L., Kizawa, K., & Ronchi, E. (2021). Interoperability of privacy and data protection frameworks. *OECD Going Digital Toolkit Note*, 21. Available at: http://goingdigital.oecd.org/data/notes/No21_ToolkitNote_PrivacyDataInteroperability.pdf.

Chapter 6

Dilemmas and Tradeoffs of Digital Governance Models: The European Union, China and United States Compared

Kayle van t'Klooster and
Madeleine O. Hosli

Abstract: How a state approaches digital governance can directly and significantly impact economic, social, and political trends domestically. The strength of democracy, especially within the context of the global democratic recession, is often argued to be closely tied to digital governance. This chapter explores how a state's chosen model of digital governance impacts the strength of its democracy over time, by exploring the different approaches to digital governance utilized by the European Union (EU), China, and the United States. The United States has primarily prioritized economic gains in its model to digital governance and while this has produced considerable economic gains, it has been at the expense of the strength of its democracy. China, on the other hand, has prioritized state control and political stability. This has indeed strengthened the Communist Party's authoritarian grip on power, but the crackdown has also stifled China's once-flourishing technology sector. Finally, the EU and its member states have developed a human-centered model that prioritizes regulation, inclusivity, and democracy. However, the EU has had issues in the past asserting its digital sovereignty and promoting its technology sector.

The chapter argues that democracy must be carefully considered when conducting digital policy. While competent digital governance can have a democratizing impact if the government embraces it fully, the opposite is also true. Often, digital governance has proven to be self-reinforcing. Governments utilize digital governance to become more democratic or authoritarian depending on their regime. However, digital governance is complex, and governments face several trade-offs and dilemmas in terms of their policy approaches.

The research question we aim to resolve with this book is: *What is the role of the EU in the digital age?*

Keywords: Digital Governance, Democracy, Relations EU-China-US, Governance Models, Trade-Offs in Digital Policies

1. Introduction

Policymakers around the world have diverged in how they approach digital governance. This chapter will summarize and analyze the three most prominent and influential approaches, the capitalist American approach, the Chinese authoritarian approach and the European regulatory approach. Each approach has costs and benefits for each respective polity. While the American approach has generated astounding economic gains for the country, it has come at the expense of the strength of its democracy. Similarly, while the Chinese government has used digital technology to strengthen the Communist Party of China's (CPC) domestic position, it has hampered the growth of its digital economy, especially beyond China's internal market. Lastly, Europe adopted a human-centric approach which seeks to uphold democratic values but risks losing out on economic activity and innovation.

This divergence in approaches to digital governance comes at a time when many democracies have weakened. Even the United States of America, once the world's most ardent defender of democracy, has not been immune. There are many causes for the democratic recession, but many academics and policymakers are increasingly blaming challenges in cyberspace as the drivers of this anti-democratic influence. Many experts have expressed concerns that advances in digital technology could weaken core aspects of democracy and democratic representation over the next decade (Rainie & Anderson, 2020). Cyberspace has become increasingly important, and its impacts non-neglectable, for democracy to reverse its decline. Moments of crisis, such as the COVID-19 pandemic or the 2022 Russian invasion of Ukraine, have underscored the importance of effective digital governance, as these crises brought with them difficult digital challenges, notably widespread disinformation.

There is a plethora of digital challenges that threaten the health of democracies globally. 2016 was an obvious watershed for this threat. Two significant votes that year, the Brexit referendum, and the American presidential election, were both colored by subsequent outrage at issues and disruptions in cyberspace (Scott, 2018; Guo et al., 2020; Frost, 2020). These high-profile votes were catalysts that spurred the increased academic attention made poor digital governance a common factor to cite when seeking to explain the recent erosion of democracy (Manville; Zuboff, 2020; Walker, 2016; Scott, 2018; Rolfe, 2016; Rainie & Anderson, 2020). Various potential undemocratic influences have been identified, such as online disinformation, political polarization and social disorder driven by online discourse and foreign influence in the democratic process (Engesser et al., 2017; Bastug et al., 2020; Frost, 2020).

While not immune to these threats, the EU is uniquely positioned to positively shape digital governance and the digital future generally in the coming years. As an established regulatory power, it has already had some successes in bringing about a more human-centric digital world which is based on democratic standards and values. However, these successes should not come at the expense of economic or technological losses. In addition, the EU must continue to work with other democratic nations, especially the United States, to ensure that there is further convergence on digital policies and the integration of democratic standards into digital technology.

Understanding the trade-offs inherent in different models of digital governance could be essential in building a knowledge base that democratic countries can employ to strengthen democracy, while minimizing losses in economic activity and innovation. If there are remedies to what ails democracy, governments in the West cannot ignore them. An analysis of the three leading approaches to digital governance helps to refine concepts and give policymakers a clear understanding of what is at stake in this debate.

2. The Chinese Approach

China's approach to digital governance has been labeled "digital authoritarianism" (Lilkov, 2020; Schneider, 2020). This approach has conceptualized digital governance as a means of strengthening state control over various aspects of society. From traditional surveillance technology like security cameras, to more dramatic measures like the now infamous social credit system, the CPC seems set on using the full extent of digital technology's more problematic capabilities. This model has successfully barricaded Chinese cyberspace from the international internet and barred its citizens from engaging with unwanted content, enabling a higher degree of state control over citizens. However, this dramatic increase in political power over the population has come at a cost. While China's tech companies like Alibaba and Tencent were once rapidly expanding with little expectation of any slowdowns in the near future, the CPC's heavy handed policy work has slowed significantly.

The basis for China's digital authoritarianism is the surveillance of its vast population by collecting tremendous amounts of data. While the largescale collection of big data is certainly not a phenomenon exclusive to China, there are three major sources which make it more effective in the Chinese market, the internet of things, smart cities and super apps. Each of these are more prevalent in China than in the West (Deng & Chen, 2018; Mozur, 2020; Wood, 2015). Abusing user data is also far from exclusively an issue in China but it is essential for the West to understand how authoritarian regimes use big data and set up protections against these sorts of practices. Unlike in the West, where data is primarily monetized through the use of hyper-targeted advertising, this data is increasingly used for pervasive surveillance. This massive amount of data has enabled Chinese authorities to make sophisticated systems such as facial-recognition systems which can identify citizens even if their faces are partially covered (Baptista, 2022).

A new policy approach, called "one person, one file" has become popular with companies and the state organizations in China. Based on this policy, new systems have been developed to organize the massive volumes of data. These systems collect data and use complex algorithms and machine learning to create customized files for individuals. According to government documents, these files even update themselves automatically as the software sorts the data (Baptista, 2022). However, just as in the West, it is China's tech giants who are collecting this data, and this has made the CPC nervous. As a result, there has been increasing domestic pressure on China's tech companies.

The CPC has engaged in a widespread crackdown on its digital technology industry, which intensified significantly in 2021. The crackdown has a number of official motivations, including the need to ensure that wealth generated by tech companies is evenly distributed among the population and the need to curb excess debt. However, it is increasingly apparent that the crackdown is inherently politically motivated (Mozur, 2020). Companies that enjoy close relationships with the government, such as Huawei, have been largely left alone while others are not as fortunate. DiDi, a once promising ride-sharing app became the target of Chinese regulators shortly after its initial public offering on the American stock exchange. Chinses regulators removed the app from Chinese app stores, arguing that DiDi was misusing user data. This resulted in Didi going from being considered the most valuable Chinese start-up to running an operating loss of $6.3 billion for the first nine months of

2021 (Yuan, 2022). The crackdown has been described by some experts as "data nationalization" and could transform this crucial economic sector into a mere appendage of the state (Yuan, 2022).

This crackdown has had negative economic consequences for the China's economy. For many tech companies, even some of the best-known tech giants, this new regulatory era is a time of layoffs, reductions and faltering growth. In the third quarter of 2021, Tencent posted its slowest revenue growth since its public listing and the e-commerce giant Alibaba's profitability declined 38 % from 2020 (Yuan, 2022). For workers, the crackdown has been similarly problematic. The extreme measures taken against the education technology and online tutoring industry, which the CPC saw as a factor driving down the birthrate, has led to millions losing their jobs. Layoffs are common across the sector, dashing the dreams of many ambitious young graduates who had hoped to join the once-prosperous industry and adding to a rising youth unemployment rate (Yuan, 2022).

Companies in China must now follow much stricter data storage rules, in pursuit of "cyber-sovereignty." Chinese companies must now ensure their data is stored domestically and there are tight restrictions on sending data out of the country. Companies which run afoul of these rules face steep financial penalties. This is part of the reason why DiDi faced such difficult regulatory blowback (Haldane, 2021). In addition, state censorship and propaganda have stifled the growth of creative outlets and platforms. iQiyi for example was once hailed as China's answer to Netflix, watched as their shared fell 85 % over the course of 2021 (Yuan). These restrictions have also slowed innovation. For some fields like AI, where access to large amounts of data, sufficient investment and a supportive environment are crucial, China is losing its international competitiveness (Yuan, 2023).

All of this domestic pressure has slowed growth in China's tech and digital sectors, but it has not killed it. More importantly for the CPC, their iron grip on political power in the world's most populous country is growing stronger. This showcases how digital governance can work as a kind of policy feedback loop. When an authoritarian country employs an authoritarian style of digital governance the impact will inevitably be a more authoritarian, perhaps totalitarian state. As seen in this example, digital governance allows a state to double down on its ideological leanings. For China and its ruling party, this means an even stronger authoritarian style which succeeds in the CPC's ultimate goal, securing its domestic position. However, China also sees digital authoritarianism as a valuable asset for exportation. The level of control China's innovation gives the state is very attractive to other authoritarian states (Wang, 2021). Therefore, it is crucial to understand China's model of digital governance so that it can be effectively countered on the global marketplace of ideas. However, when these ideological leanings differ, it is clear how digital governance can be used to achieve very different policy goals.

3. The American Approach

The American model of digital governance has privileged economic considerations above social and political implications. The ideological factor which shapes American digital governance the most is not democratic ideals, but a dedication to capitalism. A clear focus of policymakers is maintaining a laissez-faire, free-market approach, which prefers self-regulation to government regulation (Schneider, 2020). This has of course led to economic success for American businesses. The big five of Silicon Valley, Microsoft, Alphabet, Facebook, Amazon, and Apple make up a large portion of economic growth in the United States and four out of the five are worth more than $1 trillion. However, this economic success has not come without tradeoffs, which have taken the form of negative social and political trends (Pitt et al., 2018, p. 39). Indeed, the United States has become a posterchild for the global democratic decline. Placing the blame for this democratic decline on challenges in cyberspace has become a common narrative in both academic literature and popular media (Manville, 2019; Zuboff, 2020; Walker, 2016; Scott, 2021; Rolfe; Rainie & Anderson, 2020). This narrative has been further strengthened by the shocking storming of the United States Capitol building on January 6th, 2021, caused in part by online radicalization and the emergence of fringe movements like QAnon. These issues have sparked public debate on regulating American social media (Ghosh, 2021; Rubin et al., 2021; Sunstein, 2000).

The American approach, characterized by libertarian values, and a free market approach to digital governance, is argued to prevent government influence over online discourse and protect freedom of speech (Schneider, 2020, p. 3). This orientation has made private companies very wealthy but also hugely influential as companies can now determine what is acceptable content and can shape online political discourse (Zuboff, 2020). Public discourse, especially online, is increasingly problematic in the United States as a myriad of issues have arisen, such as online disinformation, political polarization, and radicalization (Engesser et al., 2017; Benkler et al., 2018; Bastug et al., 2020; Frost, 2020). Each of these issues have proven difficult for the American model to address adequately, due to its economic structure and overreliance on market solutions. The negative social impacts of social media are caused by the same features which keep costumers coming back to the platforms. These worrisome impacts are not overlooked by the public either, as a growing and clear majority of Americans agree that social media is dividing the country (Murray, 2021).

American citizens have become politically polarized, which creates antagonisms and a sense that political disagreements are increasingly personal (Pew Research Center, 2014). The average Republican and Democrat have moved further right-wing and left-wing, respectively. Even more concerning is that antipathy is surging in both groups and a growing number of both party supporters see the other party as a threat to the United States (Pew Research Center, 2014). This political discontent is partly caused by America's social media and online political discourse as social media platforms like Facebook and Twitter are prone to creating echo chambers. Citizens see specialized content that reinforces their worldviews, even if they are highly problematic or based on false information (Guo et al., p. 235). Social media companies are economically incentivized to show users content that is specially selected for their viewing. This process keeps users engaged and ensures that everything users see is based on their expressed ideological views without critical consideration (Freelon & Wells, p. 147). The only opposing views they see are often extreme and meant to cause outrage, which keeps the users engaged by manipulating their emotions and polarizes them further from other political views (Dawson, 2020, p. 67). Gradually, users become more set in their ways and potentially hostile to those who challenge them.

As Sunstein (2000) argues, citizens who have been collected into homogenous groups gradually adopt increasingly extreme views. Groups that users join on Facebook and other social media sites function like this. Sunstein (2000) also notes that including dissenting opinions can ward off polarization and extremism, but users will often moderate these groups and exclude anyone who dissents. In addition, citizens have also become more prone to being swayed by populists who capitalize on the polarization and alienation caused by social media (Engesser et al., 2017;

Rolfe, 2016; Bobba, 2018). Constant re-enforcement of one's ideological perspective can contribute to radicalization. Online radicalization is more influential in the radicalization process than traditional sources of radicalization like friends, family, radical clerics, or political figures (Bastug et al., 2020, p. 631).

Radicalization is also linked closely to exposure to disinformation and false information. The United States government has effectively allowed for its digital communications to become an incubator for disinformation, although disinformation can also originate abroad. The impact of disinformation may play a central role in America's democratic decline (Benkler et al., 2018; Lange, 2020). However, combating disinformation and other toxic content is not as simple as merely removing problematic content. This would limit public debate and establish a context in which all content must be approved by the ruling power, stifling debate and limit the space for dissenting opinions (Frost, 2020, p. 16). Despite its flaws, this is the current system. Mark Zuckerberg has said that he does not want tech companies to become "the arbiters of truth" (Fernandez, 2020). However, it is currently up to these companies to combat disinformation on their own. This responsibility also makes these companies unjustly influential in the democratic process. This fact is not lost on wealthy elites either and increasingly many see the potential for using this influence for their own benefit. For example, some have speculated that Elon Musk's interest in Twitter is motivated by the political influence it will give him (Leparmentier, 2022).

Disinformation spreads more easily in the United States due to its basic economic structure. Because disinformation can keep users engaged, digital platforms are incentivized to develop algorithms which build psychological profiles citizens and micro-target with disinformation which will keep them engaged. This function was precisely what allowed Cambridge Analytical to be so successful at influencing the 2016 American election and the Brexit referendum (Cadwalladr, 2018). This microtargeting is the business model of most American technology companies, although it is typically used for corporate advertising, not political manipulation. Zuboff (2020) describes the economic structure of American cyberspace as surveillance capitalism. Surveillance capitalism is an economic system in which massive amounts of personal data are fed into algorithms and machine learning programs to create predictive products, essentially collections of assumptions on our individuals will act in the future. These behavioral predictions are then bought and sold on markets. This system is the business model of most large American tech companies, particularly its social media platforms, and has first built and implemented by Google (Zuboff, 2020, p. 8).

Improved strategies must be implemented to counter the threat of disinformation. In the current state, the political polarization and mass manipulation of citizens through disinformation has "strikingly limited" the American public's viability as a democratic check (Graham & Svolik, 2020). Building new collaborative tools to counter disinformation, which shares the responsibility between the private companies, government, and civil society groups, would be ideal. Unfortunately, the American prioritization of economic gain and free-market policies has resulted in a system that promotes the spread of disinformation, rather than combats it. Efforts to counter disinformation may also be undercut by the financial incentive for social media companies to continue to allow problematic content. Content that is attention-grabbing and triggers an emotional reaction in the user is precisely the kind of content social media news feeds are meant to show users, since it maximizes engagement (Dawson, 2020, p. 67). For this and other reasons, social media companies are not trusted by the public to be objective in their efforts against disinformation. Many Americans, especially Republicans, believe that social media companies are purposely censoring their political views. This loss of confidence nullifies the attempts by social media companies to correct disinformation (Vogels et al., 2020).

Ultimately, the United States must reform the economic structure of cyberspace. Although it has been hugely successful in generating wealth, the current structure is unable to tackle the challenges of political polarization, radicalization and disinformation. There is also a real risk of these systems enabling mass manipulation for either corporate profit or political power (O'Connor & Weatherall, 2019; Zuboff, 2020; Dawson, 2020). The democratic woes do not come from glitches but from the system working as it was designed to.

4. The European Approach

In contrast to the Chinese and American models, the EU attempts to promote a human-centric approach to digital governance which emphasizes democratic norms and values. Some of the most prominent focuses have been privacy regulations, anti-monopoly policy and curbing the influence of large, non-European tech companies. Although previously national government had taken the lead on making digital policy, the center of gravity has shifted considerably towards the European level. The EU has taken an active role in implementing tighter regulations on user data and attempting to limit the influence of large tech companies (Goodman, 2020). Although national governments have not stopped engaging with digital governance altogether, there does seem to be some agreement that working at the EU level ensure the policies are as impactful as possible (Mărcuţ, 2020, p. 59). Specifically, it is the European Commission that has become the central actor in developing European digital policies (Mărcuţ, 2020, p. 66).

One of the first major policy proposals in the EU was the implementation of the Digital Single Market in 2015. This policy initiative was based on three pillars: access, environment and growth. Access referred to measure meant to enable better access to digital goods and services for citizens and business across the EU. Environment in this case refers to the efforts to bring about the appropriate conditions and a level playing field for digital economic activity. Lastly, the third pillar is maximising the growth potential of the European digital economy (European Commission, 2015) Generally, this policy has been well received within Europe and is found to have contributed significantly to economic growth (Marcus et al., 2019). However, its most significant effect has been the unification of the EU's significant market power into one bloc. This elimination of national boundaries in the digital economy, coupled with a European Commission which has stepped up as a coordinating actor to implement digital governance policies has enabled the EU to become a global regulatory power due to its utilization of the European market power.

The EU, directed by the European Commission chiefly, has undertaken a number of high profile and influential policy measures to govern digital technology. The most well-known and influential of these digital policies is by far the General Data Protection Regulations (GDPR). The GDPR was an ambitious piece of legislation at its inception, as it covered the data of all EU residents, regardless of where that data is processed (Goddard). Once implemented, it pushed to establish higher standards of transparency concerning how the data was used and accountability which ensured that the data was sued responsibly. Because of the wide jurisdiction the GDPR has, European regulators have often cooperated with third parties to govern big data around the world (Goddard). This cooperation with third parties means that Europe's regulatory influence in the digital sphere is significant. Many countries reference the GDPR when producing their own data protection regulations and due to the GDPR's wide scope many Chinese or American companies already have to follow the EU's regulation (Goddard, 2017; Li et al., 2019). As of 2021, the European Commission has recognized a number of countries that had adequate level of data protection in line with the GDPR, including Japan, South Korea and Argentina (European Comission, 2021).

Another high-profile plank of the EU's regulatory model is its continued struggle to enact anti-trust actions against large, often-times American tech companies. There is an observable tendency of modern high-tech companies to establish monopolies quickly and effectively. This stems from the fact that tech companies often operate based on self-reenforcing network connections, which benefit greatly from rapidly becoming the most popular and remain dominant because they are. In addition, tech companies can scale up very quickly to serve a large user base far faster than a company that relies on traditional infrastructure, especially expenses like retail space or a large workforce (Bartlett, 2018, p. 132). Tech monopolies present in developed countries today are more challenging to break up than past monopolies because anti-trust laws in many countries specifically call for action when raising the prices for consumers. Tech companies usually push the prices down for consumers, at least initially. In addition, most of the political debate on monopolies will occur on those platforms, which politicians and civil society are also now dependent on to reach their communities (Bartlett, 2018, p. 132).

The EU has shown increasing willingness to work to curtail monopolies or what it calls "gatekeeper companies," companies which have platforms on which users success other content or products. The Digital Markets Act, passed in March 2022, takes aim at companies which the EU claims use anti-competition tactics to cement their market positions. Among the companies most likely to be impacted are Google, Meta, Apple and Amazon which smaller European companies like Booking have had accommodations written into the law (Stolton, 2022). This is, however, only the latest in a long string of battles, often playing out in courtrooms in drawn out disputes, which EU competition chief Margrethe Vestager has fought. In 2017, Commissioner Vestager led the charge to levy a €2.42 billion fine on Google for abusing its dominant position in the search engine market by giving advantages to its own shopping service over competitors (European Commission, 2017). Cases like this are increasingly common in the EU and show that while the American model implicitly supports the creation of massive, influential companies which can maximize their profits, the EU would rather create an equal playing field for all.

Unfortunately, this focus on regulation and attempts to limit the influence of private companies comes with a tradeoff. The EU needs to improve how it stimulates innovation and supports growing technology companies to ensure it has a private sector which can compete with China and the United States. Some commentators argue that the EU's regulatory fixation stands in the way of innovation, just as China has similarly stifled its innovation. The development of so-called "digital champions," large American-style tech giants, has not occurred in the EU with the exception of ASML in the Netherlands. Some have argued that this is a failure of the Digital Single Market, which deprives European tech companies of the large domestic market and sources of user data which American companies benefit from more readily (Gaskell, 2019). Across the European market, many companies struggle to expand beyond their domestic borders. Increasingly many argue that the EU is overregulating its digital economy, a sector which is indeed much less explosive than the American or Chinese counterparts. There is indeed a risk that prioritizing issues like user data protections and localization over economic growth for too long may mean that the EU will be continually unable to innovate and reliant on foreign technology (Lewis, 2022).

The EU should be aware of the economic trade off which this regulatory approach may have imbedded within it. However, even though the claims that this approach is bad for business are not exaggerated, the EU still has reason to follow this regulatory framework. Many of the democratic challenges which are present in the American model of digital governance are addressed by the European model. The EU is also the only major model of digital governance which explicitly aims to safeguard liberal values and democracy over all other concerns. This is quite significant and will likely motivate the EU to stay the current course. There will likely be more efforts in the coming years to build on the EU's previous work by pushing forward with more alignment with like-minded countries on issues like devising trustworthy artificial intelligence, managing markets to balance innovation and fairness, and data to establish user control over digital identities.

5. Conclusion

Democracy must be a priority when policy makers are considering how to regulate cyberspace Europe and the United States use many of the same devices, platforms, and infrastructure components, although the majority of these are American. Building on this interoperability with further collaboration in the digital domain, with an emphasis on democracy, would benefit both parties. Each party has proven to be very adept at what the other struggles in. While the United States has generated enormous wealth form its digital sector, the EU has failed to germinate any digital champions. Simultaneously, while the United States has largely shied away from regulation in its approach to digital governance, the EU has implemented ambitious policy programs which have been influential globally. With an eye on the risk emanating from China's digital authoritarianism, The EU and America are critical to safeguarding democracy. There must be further cooperation and convergence between the two, each adopting each other's strengths, in order to find success in the future. In short, a more unified democratic model of digital governance should be developed which is committed to democratic norms and ideals, while encouraging economic growth.

Digital governance is a reinforcing factor that can either improve democracy when done right or works to accelerate a country's democratic decline if neglected. Digital governance is not an island apart from other aspects of governance but an integral part of governing a modern state. Democracies must actively decide how to govern cyberspace or face the consequences of its neglect. This is phenomenon which China understands very well, and it is precisely why the Chinese model of digital governance is designed as it is. Lilkov (2020) and Wang (2021) both show that China strengthens the power of the Chinese Communist Party through digital authoritarianism, further cementing their totalitarian rule. The democracies of the world must counter this strongly, especially as the influence of the authoritarian model creeps across the globe.

Digitalization has sown political polarization, radicalization and disinformation. Many of the major issues which are caused in part by the flaws in the American model can also be seen plainly in the EU and beyond. However, if policymakers do not deftly maneuver past the obstacles and pitfalls of cyberspace, the democratic recession will only continue to worsen. Democracy has indeed lost momentum internationally and is suffering from a kind of crisis of confidence. However, if democracy is prioritized when policymakers engage in digital governance, democracy can recover.

The EU, along with the United States, must double down to its efforts to use digital governance as a means of safeguarding democracy. This should be down by sticking with the strong stance the EU has already taken on regulation in the digital domain and continuing efforts to reach out to third parties to converge on regulations. In addition, more innovative solutions, inspired from the American model, should be employed to help grow the European technology sector. Policymakers must recognize the current tradeoffs inherent in the models of digital governance and work to correct them, so that democracy can continue to thrive in the digital age.

Bibliography

Anderson, M. & Kumar, M. (2019). Digital divide persists even as lower-income Americans make gains in tech adoption. *Pew Research Center*. Retrieved from https://www.pewresearch.org/fact-tank/2019/05/07/digital-divide-persists-even-as-lower-income-americans-make-gains-in-tech-adoption/.

Baptista, E. (2022). China uses AI software to improve its surveillance capabilities. *Reuters*. April 8. Retrieved May 15, 2022, from https://www.reuters.com/world/china/china-uses-ai-software-improve-its-surveillance-capabilities-2022-04-08/

Bartlett, J. (2018). *The People vs Tech: How the internet is killing democracy* (and how we save it). London: Ebury Press.

Bastug, M., Douai, A., & Akca, D. (2020). Exploring the "Demand Side" of online radicalization: Evidence from the Canadian context. *Studies in Conflict & Terrorism*, 43(7), 616–637. https://doi.org/10.1080/1057610X.2018.1494409

Benkler, Y., Faris, R., & Roberts, H. (2018). *Network propaganda manipulation, disinformation, and radicalization in American politics*. Oxford University Press.

Bobba, G., Cremonesi, C., Mancosu, M., & Seddone, A. (2018). Populism and the gender gap: Comparing digital engagement with populist and non-populist Facebook pages in France, Italy, and Spain. *The International Journal of Press/Politics*, 1(23).

Cadwalladr, C. (2018). "I made Steve Bannon's psychological warfare tool": Meet the data war whistleblower. *The Guardian*. March 18. Retrieved from https://www.theguardian.com/news/2018/mar/17/data-war-whistleblower-christopher-wylie-faceook-nix-bannon-trump

Colliander, J. (2019). "This is fake news": Investigating the role of conformity to other users' views when commenting on and spreading disinformation in social media. *Computers in Human Behavior* (97), 202–215. https://doi.org/10.1016/j.chb.2019.03.032

Dawson, J. (2020). Microtargeting as information warfare. *The Cyber Defense Review*, 6(1), https://doi.org/10.31235/osf.io/5wzuq

Deng, I., & Chen, C. (2018). The wechat story: How Tencent's super-app changed China. *South China Morning Post*. August 16. Retrieved April 12, 2022, from https://www.scmp.com/tech/article/2159831/how-wechat-became-chinas-everyday-mobile-app

Engesser, S., Fawz, N., & Larsson, A. (2017). Populist online communication: Introduction to the special issue, information. *Communication & Society*, 20(9), 1279–1292. https://doi.org/10.1080/1369118X.2017.1328525

European Commission. (2015). *A digital single market for Europe: Commission sets out 16 initiatives to make it happen*. Retrieved from https://ec.europa.eu/commission/presscorner/detail/en/IP_15_4919

European Commission. (2017). *Antitrust: Commission fines Google €2.42 billion for abusing dominance as search engine by giving illegal advantage to own comparison-shopping service*. Retrieved from https://ec.europa.eu/commission/presscorner/detail/en/IP_17_1784

European Commission. (2020). *Shaping Europe's digital future*. Retrieved from https://ec.europa.eu/info/strategy/priorities-2019-2024/europe-fit-digital-age/shaping-europe-digital-future_en

European Commission. (2021). *Adequacy decisions*. December 17. Retrieved from https://ec.europa.eu/info/law/law-topic/data-protection/international-dimension-data-protection/adequacy-decisions_en

Fernandez, M. (2020). Mark Zuckerberg: Social networks should not be "the arbiter of truth". *Axios*. May 28. Retrieved from https://www.axios.com/mark-zuckerberg-fact-check-twitter-facebook-0e3d9e92-00cd-4467-9ac8-bc50d4f8800e.html

Freelon, D. & Wells, C. (2020). Disinformation as political communication. *Political Communication*, 37(2), 145–156. https://doi.org/10.1080/10584609.2020.1723755

Frost, C. (2020). The power of voice: Bots, democracy, and the problem of political ventriloquism. *Journal of Political Power*, 13(1), 6–21.

Gaskell, A. (2019). Is the single market preventing digital champions in Europe? *Forbes*. September 23. Retrieved from https://www.forbes.com/sites/adigaskell/2019/09/20/is-the-single-market-preventing-digital-champions-in-europe/

Ghosh, D. (2021). Are we entering a new era of social media regulation? *Harvard Business Review*. January 14. Retrieved from https://hbr.org/2021/01/are-we-entering-a-new-era-of-social-media-regulation

Goddard, M. (2017). The EU general data protection regulation (GDPR): European regulation that has a global impact. *International Journal of Market Research*, 59(6), 703–705. https://doi.org/10.2501/ijmr-2017-050

Goodman, M. P. (2020). Time to align on digital governance. *CSIS*. Retrieved from https://www.csis.org/analysis/time-align-digital-governance

Graham, M. H. & Svolik, M. W. (2020). Democracy in America? partisanship, polarization, and the robustness of support for democracy in the United States. *American Political Science Review*, 114(2), 392–409. https://doi.org/10.1017/s0003055420000052

Guo, L., Rohde, J., & Wu, H. (2020). Who is responsible for Twitter's echo chamber problem? Evidence from 2016 U.S. election networks, information. *Communication & Society*, 23(2), 234–251. https://doi.org/10.1080/1369118X.2018.1499793

Haldane, M. (2021). What China's new data laws are and their impact on Big Tech. *South China Morning Post*. September. Retrieved from https://www.scmp.com/tech/policy/article/3147040/what-chinas-new-data-laws-are-and-their-impact-big-tech.

Kharpal, A. (2022). Tencent "exploring" a financial holding company for WeChat pay if Chinese regulators require it. *CNBC*. March 23. Retrieved May 15, 2022, from https://www.cnbc.com/2022/03/23/chinas-tencent-exploring-a-financial-holding-company-for-wechat-pay.html

Lange, E. (2020). How one social media app is beating disinformation. *Foreign Policy*. November 23.

Lee, J. (2022). The connection of everything: China and the internet of things. *MERICS*. Retrieved from https://merics.org/en/report/connection-everything-china-and-internet-things.

Leparmentier, A. (2022). Elon Musk's very political Twitter buyout sparks debate on freedom of speech. *Le Monde*. Retrieved from https://www.lemonde.fr/en/economy/article/2022/04/26/elon-musk-s-very-political-twitter-buyout-sparks-debate-on-freedom-of-speech_5981665_19.html.

Lewis, J. A. (2020). Has Europe lost both the battle and war over its digital future? *Center for Strategic and International Studies*. March 9. Retrieved from https://www.csis.org/analysis/has-europe-lost-both-bat tle-and-war-over-its-digital-future

Li, H., Yu, L., & He, W. (2019). The impact of GDPR on global technology development. *Journal of Global Information Technology Management*, 22(1), 1–6. https://doi.org/10.1080/1097198x.2019.1569186

Lilkov, D. (2020). Made in China: Tackling digital authoritarianism. *European View*, 19(1), 110–110. https://doi.org/10.1177/1781685820920121

Manville, B. (2019). Will technology kill democracy—Or reinvent it? *Forbes*. Retrieved from https://www.forbes.com/sites/brookmanvi lle/2019/09/08/will-technology-kill-democracyor--reinvent-it/?sh=53472 5d36a1b

Mărcuț, M. (2020). The governance of digital policies: Towards a new competence in the European Union. *Springer International Publishing AG*.

Marcus, J. S., Petropoulos, G., & Yeung, T. (2019, January). *Contribution to growth: The European digital single market*. Bruegel. Retrieved from https://www.bruegel.org/wp-content/uploads/2019/02/IPOL_STU20 19631044_EN.pdf

Mozur, P. (2020). Forget TikTok. China's powerhouse app is WeChat, and its power is sweeping. *The New York Times*. Retrieved from https://www.nytimes.com/2020/09/04/technology/wechat-china-united-states.html.

Mozur, P. (2021). The end of a "Gilded Age": China is bringing business to heel. *The New York Times*. October. Retrieved from https://www.nytimes.com/2021/10/05/business/china-businesses.html.

Murray, M. (2021). Poll: Nearly two-thirds of Americans say social media platforms are tearing us apart. *NBC News*. May 9 Retrieved from https://www.nbcnews.com/politics/meet-the-press/poll-nearly-two-thirds-americans-say-social-media-platforms-are-n1266773

O'Connor, C., & Weatherall, J. O. (2019). The social media propaganda problem is worse than you think. *Issues in Science and Technology*, 36(1). Retrieved from https://www.jstor.org/stable/10.2307/26949075.

Pew Research Center. (2014). *Political polarization in the American public.* Retrieved from https://www.pewresearch.org/politics/2014/06/12/political-polarization-in-the-american-public/

Pitt, J., Diaconescu, A., & Ober, J. (2018). Knowledge management for democratic governance of socio-technical systems. In *The future of digital democracy: An interdisciplinary approach.* Springer. https://doi.org/10.1007/978-3-030-05333-8

Rainie, L., & Anderson, J. (2020). Concerns about democracy in the digital age. *Pew Research Center.* Retrieved from https://www.pewresearch.org/internet/2020/02/21/concerns-about-democracy-in-the-digital-age/.

Rolfe, M. (2016). The reinvention of populist rhetoric in the digital age insiders & outsiders in democratic politics. *Springer Singapore.*

Rubin, O., Bruggeman, L., & Steakin, W. (2021). QAnon emerges as recurring theme of criminal cases tied to US Capitol siege. *ABCNews.* January 20. Retrieved from https://abcnews.go.com/US/qanon-emerges-recurring-theme-criminal-cases-tied-us/story?id=75347445

Schneider, I. (2020). Democratic governance of digital platforms and artificial intelligence? *JeDEM – EJournal of EDemocracy and Open Government,* 12(1), 1–24. https://doi.org/10.29379/jedem.v12i1.604

Scola, N. (2020). Inside the Ad Boycott that has Facebook on the defensive. *Politico.* March 7. Retrieved from https://www.politico.com/news/magazine/2020/07/03/activists-advertising-boycott-facebook-348528

Scott, M. (2018). Cambridge Analytica helped "cheat" Brexit vote and US election, claims whistleblower. *Politico.* March 27. Retrieved from https://www.politico.eu/article/cambridge-analytica-chris-wylie-brexit-trump-britain-data-protection-privacy-facebook/

Scott, M. (2021). Facebook's "Supreme Court" to receive new powers. *Politico.* February 25, Retrieved from https://www.politico.eu/article/facebook-supreme-court-thomas-hughes/

Stolton, S. (2022). EU negotiators agree new rules to rein in Tech Giants. *Politico.* March 25. Retrieved from https://www.politico.eu/article/eus-digital-markets-act-adopted/

Sunstein, C. R. (2000). Deliberative trouble? Why groups go to extremes. *The Yale Law Journal,* 110(1). https://doi.org/10.2307/797587

Sunstein, C. R. (2021). How government should regulate social media lies. *Bloomberg.* January 17. Retrieved from https://www.bloomberg.com/opinion/articles/2021-01-17/first-amendment-can-t-stop-u-s-from-regulating-social-media-lies

Vogels, E., Perrin, A., & Anderson, M. (2020). Most Americans think social media sites censor political viewpoints. *Pew Research Center.* Retrieved from https://www.pewresearch.org/internet/2020/08/19/most-americans-think-social-media-sites-censor-political-viewpoints/

Walker, C. (2016). The authoritarian threat: The Hijacking of "Soft Power". *Journal of Democracy*, 27(1), 49–63.

Wang, M. (2021). China's techno-authoritarianism has gone global. *Foreign Affairs.* Retrieved from https://www.foreignaffairs.com/articles/china/2021-04-08/chinas-techno-authoritarianism-has-gone-global

Wischmeyer, T. (2019). Making social media an instrument of democracy. *European Law Journal*, 25(2), 169–181. https://doi.org/10.1111/eulj.12312

Wood, A. (2015). The internet of things is revolutionising our lives, but standards are a must. *The Guardian.* March.

Yuan, Li. (2021). What China expects from businesses: Total surrender. *The New York Times.* July. Retrieved from https://www.nytimes.com/2021/07/19/technology/what-china-expects-from-businesses-total-surrender.html.

Yuan, Li. (2022). As Beijing takes control, Chinese Tech Companies lose jobs and hope. *The New York Times.* January. Retrieved from https://www.nytimes.com/2022/01/05/technology/china-tech-internet-crackdown-layoffs.html.

Yuan, Li. (2023). Why China didn't invent ChatGPT. *New York Times.* February 17. https://www.nytimes.com/2023/02/17/business/china-chatgpt-microsoft-openai.html.

Zuboff, S. (2020). *The age of surveillance capitalism: The fight for a human future at the new frontier of power.* Public Affairs.

Chapter 7

The Algorithmic Life Seen from The Americas: An Opportunity to Deploy EU Diplomacy with a Social Impact

Paulina Astroza and Matthias Erlandsen

Abstract: The EU's international aid and cooperation is an icon of commitment to universal human values and particularly to the European ones. In the context of the EU priorities 2019–2024, particularly in the so-called Digital Decade of Europe, the EU considers training in digital matters a fundamental pillar of its policy. Taking the previous experiences of technological cooperation and its financing in the 2000s, today the EU once again has the opportunity to generate spaces to help modernise Latin American diplomatic services, and democratise international affairs for the citizens, particularly for young people. This chapter discusses those opportunities.

Keywords: AI, Algorithms, Democracy, EU Diplomacy, Latin America

1. Introduction

The Americas is often behind in many aspects of social development (Fukuyama, 2008) and technological development (Bastos, 2004). However, it is one of the most advanced regions of the world in the discussion and regulation on artificial intelligence (Mazetti, 2021). After the almost unmatched power that the EU showed to regulate aspects of digital life that are becoming urgent, such as the protection of personal data online, here we reflect on the window of opportunity that the European diplomacy has in order to take the example of the Americas and, in the same way that it did with the regulation on data protection, advocate on the regulation of the quality of algorithms, so other States can protect democracies and particularly promote human rights through the concept of diplomacy with social impact.

Our work proceeds as follows, the following section examines the current situation regarding data as the new commodity, highlighting the fact that even though we do not realize, we are embedded in a world which listens and records abundant personal data from our phones and online interactions. Then, our work explains why neglecting regulation on this fact is risky. Here we also distinguish between data protection and regulation of the quality of algorithms, while we also delve into what different international actors have done about it. In the next section, our work particularly refers to the consequences on democracy and human rights if algorithms do not get somehow regulated. In the last section, we explain the window of opportunity that the EU has when addressing this issue mimicking the strategy it had in recent years with the GDPR by putting human rights and citizens in the center of the issue, for which we coin the concept of diplomacy with social impact.

2. Data Are the New Commodity

The global economy and global power is changing. If only thirty years ago oil and nuclear weapons were indicators of power, that is now data and information.

Data—which shapes information—such as your routine around the city, or your shopping list is what both private companies and governments seek, because this way they can know you better and enter your mind. Yes, you read it right—seep into your mind.

Has it ever happened to you that you often think about a product, or you comment about it very quickly to the person next to you at that moment, and a few minutes later, while browsing the feed of a social media platform, an advertisement appears for you? That same product you thought of? We are sorry to disappoint you, but it is not a mere coincidence.

Every click or tap on our devices generates data that is stored on the Internet. We do not realize it, but each person who has a smartphone or a computer connected to the web creates thousands of data that they unconsciously store somewhere and to which they rarely have direct access or even control over them. The amount of data stored online is incredibly huge. Humbetov (2012) claims that 90 % of the data has been created only in the past two years—and it keeps growing.

The data is highly desired by private companies—and also by governments. The formers, because it allows them to make decisions through artificial intelligence mechanisms and sell increasingly specific products or services, generating higher profits and lower losses through the use of this new unregulated technology (Brown et al., 2021); and the latters, because although personal information could be used by governments to design public policies, they also serve to closely monitor citizens. This way, the benefits also bring us costs, because after giving personal data, this artificial intelligence works by discriminating information, products, or services that are not of our interest—at least at that specific moment—and makes that selection last over the time, generating biases (Sadowski et al., 2021). This type of artificial intelligence is called algorithms.

According to Pauli (2018) "An algorithm is a finite sequence of instructions for performing a task". In Dourish's words (2016, p. 3), "it is an abstract, formalized description of a computational procedure". In a simpler way, algorithms are the masterminds that feed and run many of the softwares that control social media and e-commerce platforms that learn from our choices over the time, and that are able to make autonomous decisions. In social media, algorithms help maintain order and assist in ranking search results and advertisements. On Facebook, for example, there is an algorithm that directs pages and content to display in a certain order, that is why, for example, you see ads about coffee in the mornings or last minute quick short trips on Fridays' afternoon.

As we mentioned, these algorithms are built and trained while a person uses and consumes social media. The more anybody uses a social platform, the more data the algorithm gets, and the better it can be fine-tuned. Now, guess what is the real problem? Exactly: kids and young people's algorithms.

Unlike those of us who were born before the 2000s, the centennial generation—and those that followed—no matter where in the world, they were born and grew up with mobile screens always connected to the Internet. Companies like Google, Facebook, or Amazon have had much more time to collect and store personal data from these youngsters, so these people's algorithm becomes more specific, offering a limited variety of ideas and products, discriminating against the plurality of reality in which society develops, and enclosing these generations in informative echo chambers, which, as we will see later, can be extremely harmful. This being the case, somehow it is necessary to prevent, and diplomacy in general—but EU diplomacy in particular—has a unique window of opportunity to exercise socially responsible diplomacy, putting human rights at the center of the discussion of an issue as important as new, through its networks and lobbying power.

3. Why Legislate? The Risk of Inaction

The Internet is a difficult space to regulate (Santaniello, 2021). Because it is an intangible place that crosses borders, under what laws or rules can we organize ourselves so that these personal data are not left at the mercy of just anyone? In its beginnings, it was thought that the Internet should be a self-regulated space (Price & Verhulst, 2005), because it was assumed that its access was rather restricted to an academic, defense and security, and scientific elite (Abbate, 1999; Gillies & Cailliau, 2000; Hafner & Lyon, 1998). With the explosive and increasingly easy access that a large part of the planet's inhabitants has today to the network[1], problems such as the malicious use or the abuse of data stored on the web has emerged as a constant problem for each one of the states of the world. Furthermore, this very use and abuse of our data can contribute to the manipulation of emotions in the electorate who, without realizing it, ultimately make political decisions based on the content they are exposed to through their social media. The latter can seriously jeopardize the democratic system itself.

The literature continues to defend that there is an asymmetry of protection regarding digital data, which has an impact on societies and individuals, from different points of view.

For example, there are those who argue that the conceptual difference by which data protection is defended in Europe compared to other places—mainly based on a guarantee of Human Rights—, would imply that there are different Human Rights systems, which is opposed to its universality (Maqueo Ramírez et al., 2017). By not having a universal and unchangeable definition, it is possible that many other aspects of social life would also be affected, such as the equality and non-discrimination of people, in the sense that much of the information stored without protection can deliver data that implies the isolation, segregation and arbitrary discrimination of citizens; not legislating regarding personal digital data such as political, religious or sexual preferences, implies risks for people even at levels of personal security, because without any regulation it is not possible to guarantee freedom of expression, free association, freedom of thought, and even self-determination (Cerda Silva, 2011).

---

1 In Estonia, France and Greece, Internet access is considered a fundamental right of every human being (Tully, 2014).

Moreover, as we anticipated, we could claim that the risk of not properly legislating on the protection of digital data can put democracies at risk. We just need to remember the United States electoral campaign of 2016, when Donald Trump was elected president, who for the first time in history hired a strategist and director of social media. Brad Pascale, the person in that position, later assured that the command of the now former president had data of at least 50 million of voters, which in turn helped them create specific content that could have played a central role in the results of said race for the presidential chair.

This same discussion was held in 2016 after the British referendum on the permanence of the United Kingdom in the European Union (Gorodnichenko et al., 2021), and again in 2018, during the presidential elections in Brazil in which Jair Bolsonaro was elected (Santos, 2020), and again around the presidential election in Russia that same year (Gunitsky, 2020). Are the results of these elections really the will of the people or is it rather a manipulation of unprotected or poorly regulated data, which ends up being used strategically in favor of a particular candidate?

In Europe, the issue of regulation of citizens' personal information was discussed and settled around the General Data Protection Regulation (GDPR) Directive of 2016. In its spirit, the GDPR seeks to update and comply with the law of the Right to the Privacy of 1950 that emanates from the European Convention on Human Rights, so that everyone has the "right to respect for their privacy and their family life, their home and their correspondence". The GDPR is one of the most important and well-known regulations on the matter and has become the model for other States to legislate in this regard. Notwithstanding, the GDPR says nothing about algorithms.

At this point we must differentiate between the legislation regarding data protection—where the EU is the world leader (Law Library of Congress (U.S.), 2019)—, that is, regulations and safeguardings of the information that we have "voluntarily" provided to a company or State, versus the regulation around the quality of the algorithms that are built on different digital platforms thanks to the data we give (Lyubashits et al., 2015), and that are closely linked to neuro-rights and artificial intelligence—where Latin America, but specifically Chile, are ahead (Diaz Moraga, 2022; Girardi Lavín et al., 2022). This loophole in the Americas is a perfect place to be influenced by other international actors with pretensions and geopolitical influence in the region—namely the United States and China.

Moreover, Bauzá Martorell (2019) clearly makes the point when stating that, in terms of data protection, there are two clearly differentiated conceptions: the European one—focused on privacy—and the American, which subordinates privacy to security. He also points out that Latin American countries have advanced towards a process of legislating on personal data protection regulations, closer to the European model than to the American one, and consequently they have a special sensitivity for privacy.

After the Facebook Papers in the fall of 2021, and the testimony of Frances Haugen in the US Congress about how the company uses algorithms to target users with specialized content triggering effects over political and societal issues, the Capitol experienced an extraordinary number of bills on the regulation of algorithms, adding up over 31 projects by 2021. Notwithstanding, according to Samp, Phillips and Tobey (2022), the sentiment among lawmakers in the short run appears to use the existing structures, instead of creating an entirely new regulatory architecture from scratch. For example, one of the most discussed ideas in this regard is a reform to Section 230 provision in the 1996 Communications Decency Act, which provides immunity for website platforms that publish information from third-party content.

Yet, the real problem about legislating on algorithms in the United States by amending this particular Section is the fine line between enforcement against certain content and the free speech concern. The debate around the protection of democracy or human rights—aside from the First Amendment—, and whether this new regulation would also apply to the US government or other actors overseas does not seem to be on the table so far.

Across the Pacific, China established the world's most ambitious effort to regulate artificial intelligence in March 2022, known as the Internet Information Service Algorithmic Recommendation Management Provisions, drafted by the Cyberspace Administration of China.

The regulation, which seeks to outlaw algorithmic bias and discriminations, will also prohibit companies from using personal information to offer users specific products or services according to their personal characteristics or preferences at a different market value—the same product, for the same price, for everybody—. At the same time, these new rules specifically aim at regulating algorithms that control results, filter contents among the main e-commerce, streaming, social media platforms, and car-sharing companies.

These legislations go one step ahead because in the recent past China has ruled on banning the ranking of celebrities by popularity on the main social media platforms, and called those companies to control the access to underage citizens by not inducing them to consume or incite to buy something in order to vote in TV shows; law that complies with the regulation on the limits to only one hour—only on Fridays, weekends and public holidays—on the time that Chinese children could spend playing online video games, affecting about 110 million under the age of 18.

Also, the Cyberspace Administration of China (2022) presented a proposal to legislate around what the government calls "synthetic content", a term that includes fake news, and deepfake audiovisual materials made by blending two persons' faces together using artificial intelligence. Gabriele de Seta (2021) argues that in contrast to the EU regulations that tend to think in terms of fostering markets and individuals' rights, in China it is all about societal wellbeing. All in all, as with the bills in the US, in the case of China it is not yet clear what the scope would be to the State, nor the impact it would have on companies and citizens based abroad.

Across orient all the way to Europe, although it is true that with the entry into force of the GDPR many companies around the world had to adapt—the GDPR is applicable to any organization outside the block that offers goods or services (paid or free) or that monitors the behavior of individuals from the EU—, the current state of the legislation is unbalanced, because little has been discussed on the regulation of algorithms affecting citizens' decisions.

That is not the case of the Americas, because while countries like Argentina or Uruguay are recognized for their high level of concern and detail in the matter of artificial intelligence, other States such as Bolivia, El Salvador, and Guatemala, do not have regulations on any digital life topic. Furthermore, comparative research often leaves smaller Caribbean countries out of its analyses.

Yet, as we saw, data protection and regulation of algorithms are two different topics that need to be addressed simultaneously.

4. An Unregulated Algorithmic Life

Tech businesspeople such as Intel CEO Brian Krzanich, and other academics in the business and management fields, such as Kaplan and Haenlein (2019) argue against the regulation of artificial intelligence and algorithms, claiming that technology in this area is still too young and under development, yet they suggest common norms including requirements for the testing and transparency of algorithms, possibly in combination with some form of warranty.

On the contrary, some scholars argue in favor of regulating algorithms, both in the ethical (Tsamados et al., 2022) and in the legal fields (Cai & Ji, 2020; Brunetta et al., 2022), because—they say—on the one hand, artificial intelligence could replicate existing social inequalities along race, class, gender, and sexuality lines (Buolamwini, 2022; Fitsilis, 2019; Ledford, 2019); and on the other hand, some academics—mainly in communications (Garrett, 2009; Zimmer et al., 2019), and political science (Sunstein, 2001)—pay attention to legislation in order to avoid echo chambers built inside filter bubbles (Pariser, 2011) that in turn are a consequence of algorithms, which is this

> "state of intellectual isolation determined by the preference algorithms that underlie contemporary web-based platforms such as Facebook and Google, [...] be[ing] limited to interact and consume information in an "ideologically safe" and unchallenged environment due to preference algorithms." (Figà Talamanca & Arfini, 2022).

Of course, this concern has been discussed amongst the academics in technologies: we cannot miss the opportunity to note that back in 1942, Isaac Asimov already had addressed the issue on the regulation of algorithms in his short story "Runarounds" (1950), where he states the "Three Laws of Robotics". Closer to the present, Sir Tim Berners-Lee—the inventor of the World Wide Web—back in 2017, for the 28th anniversary of the Internet, called to tackle three challenges that we have in order for the web to fulfill its true potential as a tool which serves all of humanity, being the first the loss of our personal data, and the second, the easy spread of misinformation on the web based on algorithms which learn from our personal data that online companies are constantly harvesting.

As we mentioned, the problem lies in the protection that citizens have against the use or manipulation of that data—even more than its potential leak—, but above all, the focus should be on the quality of said algorithms (Raji et al., 2020)—built based on these data—specifically in the algorithms that condition the new generations of citizens. Neglect any legislation in this regard will inevitably have an impact on the quality of democracy, giving reason to thinkers like Peter Thiels to argue that freedom and democracy will no longer be compatible in the algorithmic society (Fukuyama, 2012; Thiels, 2009), opening the possibility to more examples such as those already experienced in the elections of Donald Trump and Jair Bolsonaro, during the Brexit, or boost the wave of populist governments that profit from manipulation, the intensive use of social media, and fake-news—or rightly lies—that jeopardize the democratic game in liberal States.

Algorithmic life mainly works to the detriment of the reflective capacity of human beings in intellectual development. Centennial generations and those yet to come grew up with screens and Internet access under very little parental control. It is in this sense, where the "filter bubbles" and "echo chambers" (Berman & Katona, 2020) will offer them a biased and small portion of reality, and it is there where the danger of misinformation puts the quality of democracies in danger, because of the reduction of the cognitive universe of the users. It would not be strange then, that in the mid-run, we inhabit a world without societies as we know them today, but instead one divided into a kind of digital castes managed by the quality and plurality of the algorithm that feeds us with information (Sadin, 2009, 2013, 2015, 2021a, 2021b; Zuboff, 2020).

5. A Window of Opportunity for the EU: Diplomacy with Social Impact

Algorithms will be widely used for decision making in many aspects of our future lives because they reduce the margin of error of the outcome, based on probabilities and statistics. There is nothing wrong with that. But, if the decision does not go through human reflection and logical analysis—similar to what we could say of the qualitative method—, there is a risk that the decision will be perpetuated and that, at the international level, for example in the case of wars, those will never end in peace negotiations, but in an escalation of violence given that the end would always be the total benefit and hegemony of only one of the parties—the stronger algorithm.

The risk for International Relations of having algorithms with unregulated quality is that decisions could be made not in order to seek the national interest but rather the interest of the owner of the algorithm. This puts security, the sovereignty of States, democracy, and of course human rights, at risk. The new battlefield is the brain, to the detriment and risk of our own biology that leaves us vulnerable to manipulation and totalitarianism.

Diplomacy would be meaningless if it did not put the interest of the State at the center of all its actions (Adler-Nissen, 2016). In addition, understanding that the interest of the State should mirror the interest and needs of its citizens—and therefore to be a reflection of the domestic policy—, then diplomacy should work aiming at generating a social impact overseas. In this sense, the fundamental and central role of European diplomacy in this matter should remain attached to its fidelity for the respect and promotion of human rights and deploy its capacities to raise awareness of the urgency about the need to legislate on the quality of the algorithms, which if not done soon, it would directly affect its citizens. This diplomacy with social impact[2]—as we have named this kind of actions—differs from propaganda, yet in a way takes some characteristics of the advocacy task that traditional definitions of diplomacy have (United Nations, 2005) because it seeks to improve the quality of life of citizens in whichever mission diplomats are deployed, while it is heavily related to the concept of "Human Rights Diplomacy" as argued by Petrasek (2013) in the sense that it no longer pursues tactics such as public criticism and conditionality, but it may embrace other tactics, including dialogue-driven approaches and thematic-specific standard-setting.

And it is in this sense that the EU has a window of opportunity, given the fact that in the recent past it has already shown that it does have the power at an international level to call on other States to legislate on data protection. It is a manifestation of their soft power (Nye, 2004). Now the EU has the opportunity to legislate on the manipulation of data and the quality of algorithms, so that other geopolitical zones need to follow suit.

Undoubtedly, European diplomacy can emphasize the protection of its interests, especially when it comes to the protection of Human Rights

---

2 We would like to thank ambassador Marina Rosenberg (from Israel) who gave us the incipient idea around this concept that we further developed.

and democracy, for example, highlighting the importance of having legislation that guarantees pluralist algorithms, and that reflect society as a whole. EU diplomacy can do this through communication and awareness campaigns towards the population, so that citizens understand the importance of having legislation that represents and protects them and thus put pressure on their parliamentarians, preventing this way that other international actors achieve greater power and influence, against a regulation focused on the preservation of a broad algorithm, which even endangers democracy and respect for Human Rights, as we have already exposed.

Strong international pressure on this matter is essential, so that no region of the planet becomes a "data haven," which are spaces similar to "tax havens" or "corporate heavens" in which poor legal regulation and little government oversight regarding privacy and the use of information becomes a friendly space for companies to benefit and profit, which in turn leads to increased levels of corruption, and many other digital crimes. Examples like these abound in Southeast Asia, Africa, and the Caribbean.

In turn, European diplomacy can lobby for laws and regulations, perhaps not as in an specific international treaty on algorithmic regulation, but through other accords, such as the case of the modernization of the Chile-European Union Association Agreement, in which the topic of data protection was addressed in the chapter on Digital Economy regulating electronic commerce, and transfer of data—although, nothing on regulation of algorithms is present in this new agreement—. This being the case, European influence, if not through the signing of an international treaty as a whole Latin American bloc, can at least influence various States as individual actors. This idea agrees with scenario number 2 argued by Crocker, Hampson, and Aall (2021) on a new world order returning to international cooperation, which Ana Palacio (2021) calls governance diplomacy and reminds us that is the core of the European Union's statecraft.

If the EU does not deploys its diplomatic force to address this issue, other international actors such as the US and China will do so, and probably in a neoliberal way, that is, without putting the citizens at the center, but rather the big companies, because it is in their interest that the region becomes a data heaven, which goes against the spirit of the EU with human rights at its core. After all, data as a new commodity is here to stay, and we will not be able to fight that change.

References

Abbate, J. (1999). *Inventing of the internet.* Cambridge, MA: MIT Press.

Adler-Nissen, R. (2016). Diplomatic agency. In C. M. Constantinou, P. Kerr, & P. Sharp (eds.), *The SAGE handbook of diplomacy* (pp. 92–103). London, UK: SAGE.

Asimov, I. (1950). Runaround. In *I, Robot* (The Isaac Asimov Collection ed.). New York, NY: Doubleday.

Bastos, M. I. (2004). Politics, the state and policies for science and technology in Latin America. In M. I. Bastos (ed.), *The politics of technology in Latin America*. https://doi.org/10.4324/9780203208755

Bauzá Martorell, F. J. (2019). El modelo europeo de protección de datos: Experiencias para la regulación chilena presente y futura. *Ars Boni et Aequi*, 15(1), 121–148. https://doi.org/10.23854/07192568.2019152Bauza121

Berman, R., & Katona, Z. (2020). Curation algorithms and filter bubbles in social networks. *Marketing Science*, 39(2), 285–457, ii. https://doi.org/10.1287/mksc.2019.1208

Berners-Lee, T. (2017). *Three challenges for the web, according to its inventor.* Retrieved from https://webfoundation.org/2017/03/web-turns-28-letter/

Brown, S., Davidovic, J., & Hasan, A. (2021). The algorithm audit: Scoring the algorithms that score us. *Big Data & Society*, 8(1), 1–8. https://doi.org/10.1177/2053951720983865

Brunetta, C., Studart Leitão, A., & Rocha Dias, E. (2022). Da pessoa e para a pessoa: A regulação jurídica dos algoritmos. *Revista de Informação Legislativa: RIL*, 59(233), 163–178.

Buolamwini, J. (2022). *Facing the Coded Gaze with Evocative Audits and Algorithmic Audits.* (Doctoral Thesis). Massachusetts Institute of Technology, Cambirdge, MA. Retrieved from https://hdl.handle.net/1721.1/143396

Cai, L., & Ji, X. (2020). Legal regulation of algorithms from the perspective of interpretability. *Cross-Cultural Communication*, *16*(1), 65–75. https://doi.org/10.3968/11612

Cerda Silva, A. J. (2011). *Personal data protection and online services in Latin America* freedom of speech and internet: Issues on regulation in Latin America. Center for Studies on Freedom of Expression and Access to Information at the University of Palermo Law School. Buenos Aires. Retrieved from https://www.palermo.edu/cele/pdf/english/Internet-Free-of-Censorship/03-Personal_Data_Protection_Online_Services_Latin_America%20_Alberto_Cerda.pdf

Crocker, C. A., Hampson, F. O., & Aall, P. (2021). Diplomacy and world order. In C. A. Crocker, F. O. Hampson, & P. Aall (eds.), *Diplomacy and the future of world order* (pp. 287–296). Washington, DC.: Georgetown University Press.

Cyberspace Administration of China. (2022). Notice of the cyberspace administration of china on public solicitation of comments on the provisions on the administration of deep synthesis of internet information services (Draft for Comment). Retrieved from http://www.cac.gov.cn/2022-01/28/c_1644970458520968.htm

de Seta, G. (2021). Huanlian, or changing faces: Deepfakes on Chinese digital media platforms. *Convergence: The International Journal of Research into New Media Technologies*, 27(4), 935–953. https://doi.org/10.1177/13548565211030185

Díaz Moraga, M. J. (2022). *Neuroderechos y datos personales sensibles: Análisis de la posible regulación superpuesta*. Concepción, Chile: Universidad de Concepción.

Dourish, P. (2016). Algorithms and their others: Algorithmic culture in context. *Big Data & Society*, 3(2), 1–11. https://doi.org/10.1177/2053951716665128

Figà Talamanca, G., & Arfini, S. (2022). Through the newsfeed glass: Rethinking filter bubbles and echo chambers. *Philosophy & Technology*, 35(20). https://doi.org/10.1007/s13347-021-00494-z

Fitsilis, F. (2019). *Imposing regulation on advanced algorithms.* Cham: Springer.

Fukuyama, F. (2008). Poverty, inequality, and democracy: The Latin American experience. *Journal of Democracy*, 19(4), 69–79. https://doi.org/10.1353/jod.0.0035

Fukuyama, F. (2012). A conversation with Peter Thiel. *The American Interest*, 7(4). Retrieved from https://www.the-american-interest.com/2012/02/01/a-conversation-with-peter-thiel/

Garrett, R. K. (2009). Echo chambers online?: Politically motivated selective exposure among Internet news users. *Journal of Computer-Mediated Communication*, 14, 265–285. https://doi.org/10.1111/j.1083-6101.2009.01440.x

Gillies, J., & Cailliau, R. (2000). *How the web was born: The story of the World Wide Web*. New York, NY: Oxford University Press.

Girardi Lavín, G., Hoelm, M., Amar, M., Vásquez, D., & Walker, J. (Eds.). (2022). *Chile tiene Futuro desde sus Territorios*. Santiago, Chile: Ediciones Biblioteca del Congreso Nacional de Chile.

Gorodnichenko, Y., Pham, T., & Talavera, O. (2021). Social media, sentiment and public opinions: Evidence from #Brexit and #USElection. *European Economic Review*, 136, 103772. https://doi.org/10.1016/j.euroecorev.2021.103772

Gunitsky, S. (2020). Democracies can't blame Putin for their disinformation problem. *Foreign Policy* (21). Retrieved from https://foreignpolicy.com/2020/04/21/democracies-disinformation-russia-china-homegrown/

Hafner, K., & Lyon, M. (1998). *Where wizards stay up late: The origins of the internet* (First Touchstone Edition 1998 ed.). New York, NY: Simon & Schuster.

Humbetov, S. (2012). D*ata-intensive computing with map-reduce and hadoop*. Paper presented at the 2012 6th International Conference on Application of Information and Communication Technologies (AICT), Tbilisi, Georgia.

Kaplan, A., & Haenlein, M. (2019). Siri, Siri, in my hand: Who's the fairest in the land? On the interpretations, illustrations, and implications of artificial intelligence. *Business Horizons,* 62(1), 15–25. https://doi.org/10.1016/j.bushor.2018.08.004

Law Library of Congress (U.S.). (2019). *Regulation of artificial intelligence in selected jurisdictions*.

Ledford, H. (2019). Millions affected by racial bias in health-care algorithm. *Nature*, 574, 608–609. https://doi.org/10.1038/d41586-019-03228-6

Lyubashits, V. Y., Mordovtsev, A. Y., & Mamychev, A. Y. (2015). *State and Algorithms of Globalization. Mediterranean Journal of Social Sciences,* 6(3 S6), 277. https://doi.org/10.5901/mjss.2015.v6n3s6p277

Maqueo Ramírez, M. S., Moreno González, J., & Recio Gayo, M. (2017). Protección de datos personales, privacidad y vida privada: la inquietante búsqueda de un equilibrio global necesario. *Revista de Derecho (Valdivia),* 30(1), 77–96. https://doi.org/10.4067/S0718-09502017000100004

Marzetti, M. (2021). Algorithmic law—A legal framework for Artificial Intelligence in Latin America. In A. Plaw, B. Carvalho Gurgel, & D. Ramírez Plascencia (eds.), *The politics of technology in Latin America: Data protection, homeland security and the labor market* (Vol. 1). Oxon: Routledge.

Nye, J. S. (2004). *The soft power: The means to success in world politics.* New York City, NY: Public Affairs.

Palacio, A. (2021). Europe's persistent gap between rhetoric and reality. In C. A. Crocker, F. O. Hampson, & P. Aall (eds.), *Diplomacy and the future of world order* (pp. 78–92). Washington, DC.: Georgetown University Press.

Pariser, E. (2011). *The filter bubble: What the internet is hiding from you.* London, UK: Penguin Press.

Pauli, S. (2018). *MAT 112 ancient and contemporary mathematics.* Retrieved from https://mathstats.uncg.edu/sites/pauli/112/HTML/section-5.html

Petrasek, D. (2013). New powers, new approaches? human rights diplomacy in the 21st century. *International Journal on Human Rights,* 10(19), 7–14. Retrieved from https://sur.conectas.org/en/human-rights-diplomacy-21st-century/

Price, M. E., & Verhulst, S. G. (2005). *Self-regulation and the internet.* The Hague: Kluwer Law International.

Raji, I. D., Smart, A., White, R. N., Mitchell, M., Gebru, T., Hutchinson, B., . . . Barnes, P. (2020). *Closing the AI accountability gap: defining an end-to-end framework for internal algorithmic auditing.* Paper presented at the Proceedings of the 2020 Conference on Fairness, Accountability, and Transparency, Barcelona, Spain. https://doi.org/10.1145/3351095.3372873

Regulation (EU) 2016/679 of the European Parliament and of the Council of 27 April 2016 on the protection of natural persons with regard to the processing of personal data and on the free movement of such data, and repealing Directive 95/46/EC (General Data Protection Regulation), 32016R0679 (2016).

Sadin, É. (2009). *Surveillance globale: enquête sur les nouvelles formes de contrôle.* Paris: Climats.

Sadin, É. (2013). *L'humanité augmentée: l'administration numérique du monde.* Montreuil: Éditions L'Échappée.

Sadin, É. (2015). *La vie algorithmique: critique de la raison numérique.* Paris: Éditions L'Échappée.

Sadin, É. (2021a). *L'intelligence artificielle ou l'enjeu du siècle: anatomie d'un antihumanisme radical.* Paris: E´ditions L'Échappée.

Sadin, É. (2021b). *La silicolonisation du monde: l'irrésistible expansion du libéralisme numérique.* Paris: E´ditions L'E´chappe´e.

Sadowski, J., Viljoen, S., & Whittaker, M. (2021). Everyone should decide how their digital data are used—not just tech companies. *Nature*, 595, 169–171. https://doi.org/10.1038/d41586-021-01812-3

Samp, T., Phillips, S. R., & Tobey, D. (2022). US Congress tries to decode algorithms. Retrieved from https://www.dlapiper.com/en/us/insights/publications/2022/1/us-congress-tries-to-decode-algorithms/

Santaniello, M. (2021). From governance denial to state regulation: A controversy-based typology of internet governance models. In B. Haggart, N. Tusikov, & J. A. Scholte (eds.), *Power and authority in internet governance: Return of the state?* London, UK: Routledge.

Santos, B. R. G. (2020). Echo chambers, ignorance and domination. *Social Epistemology, 35*(2), 109–119. https://doi.org/10.1080/02691728.2020.1839590

Sunstein, C. R. (2001). *Republic.com*. Princeton, NJ: Princeton University Press.

Thiel, P. (2009). The education of a libertarian. *From Scratch: Libertarian Institutions and Communities.* Retrieved from https://www.cato-unbound.org/2009/04/13/peter-thiel/education-libertarian/

Tsamados, A., Aggarwal, N., Cowls, J., Morley, J., Roberts, H., Taddeo, M., & Floridi, L. (2022). The ethics of algorithms: key problems and solutions. *AI & Society*, 37, 215–230. https://doi.org/10.1007/s00146-021-01154-8

Tully, S. (2014). A human right to access the internet? Problems and prospects. *Human Rights Law Review*, 14(2), 175–195. https://doi.org/10.1093/hrlr/ngu011

Vienna Convention on Diplomatic Relations 1961, (2005).

Zimmer, F., Scheibe, K., Stock, M., & Stock, W. G. (2019). Fake news in social media: Bad algorithms or biased users? *Journal of information Science Theory and Practice*, 7(2), 40–53. https://doi.org/10.1633/JISTaP.2019.7.2.4

Zuboff, S. (2020). *The age of surveillance capitalism: The fight for a human future at the new frontier of power.* New York, NY: Public Affairs.

Chapter 8

Transnational Skills Exchanges as a Model for Science Diplomacy: European Aid and Reconstruction in Palestine

JONATHAN MICHAEL FELDMAN

Abstract: This chapter will examine an initiative led and organized by a Palestinian knowledge and policy entrepreneur based in Sweden to promote a biomedical laboratory in Palestine as well as the larger mechanisms supporting related Swedish-Palestinian science exchanges. I argue that this and related initiatives are a potential model for EU science policy in the digital era. Such projects can support diverse aims including economic aid support, health security, peacebuilding and alliance development. Key enabling mechanisms include actions by academic entrepreneurs, state foreign and aid policy, and diasporic networks. Future EU policy could learn from and replicate the initiatives discussed.

Keywords: Citizen diplomacy, innovation, Palestine, regional development, transnational skills exchanges

1. Introduction

This chapter will examine initiatives designed to promote science diplomacy between Sweden and Palestine involving transnational skill exchanges. I have two primary research questions. My first question is: "To what extent do university-based scientific exchanges constitute science diplomacy?" Here I am interested in exploring how university exchanges can constitute a kind of *science diplomacy from below*. My second research question is: "How can transnational skills exchanges promote various goals such as health security or diplomatic goals and thereby serve as a model for EU science policy?" Here I examine potential benefits from university exchanges and how they can contribute to these stated diplomatic goals.

I will answer these questions by examining theories of Palestinian development, diasporic linkages, university exchanges and organizational commitments and science diplomacy. After providing a background to Palestinian developments, I then examine the case of a Swedish Palestinian knowledge and policy entrepreneur who organized various exchange programs. I argue that there are potential bases for more systemic reconstruction of Palestinian society, if policy makers extend key enabling mechanisms including: actions by academic entrepreneurs, state foreign and aid policy, and diasporic networks. Future EU policy could learn from and replicate the initiatives discussed.

2. Background

Under what conditions can innovation, science and technology serve democratic goals? This question has long been contested with some observers looking at the negative aspects and others emphasizing the positive. The question can only be answered by examining the social context under which technologies are promoted. The case of European foreign relations with Palestine provides us with a very useful test for ideas related to technology, foreign aid and diplomacy. The politics of technology transfer in the context of foreign aid to an occupied territory raises many difficult questions. These center on whether technologies, growth and aid help legitimate or consolidate such occupations or provide a means of surviving, negotiating or potentially helping to transcend them. The reality may be somewhere in between.

Palestine presently suffers from various inter-related problems such as high levels of unemployment, occupation by Israel and constraints on democratic development. A key question has been how to advance Palestinian development in a region that is constrained by occupation and a severe security regime. Various frameworks interpret this regime as triggered by "settler colonialism" (where the focus is on Israeli actions and responsibilities) or fundamentalist extremism (with the focus is on Palestinian actions and responsibilities) (Israeli, 2002; Lloyd, 2012). Some scholars also see a connection between extremism on the one hand and occupation and or the severe security regime on the other (Hafez, 2006; Hsu & McDowall, 2020; Pape, 2006).

With this background the question that emerges is the extent to which science diplomacy might play a role in ameliorating conditions in Palestine. This paper addresses the question of how science diplomacy from below could intervene in development through academic entrepreneurial efforts despite constraints. Such diplomacy can promote capacity development, health security, peacebuilding and alliance development.

3. Theory

3.1. Introduction

Various theories address the problem of development against the backdrop of military occupation. Theories of Palestinian development explore internal and external constraints on the expansion of capacities if not growth. One way to expand capacities in a peripheral region is to through linkages to core regions organized by diasporic networks. Another way is through university efforts in the core on behalf of counterparts in the periphery. I show how such capacity extension systems can be characterized as a form of science diplomacy in the concluding subsection.

3.2. Palestinian Development

One argument is that economic prospects for Palestinians are limited by "the absence of sovereignty." Development depends on the ability to have a sovereign nation state, so that "economic growth" is tied "to the growth of national capabilities and not simply to rising standards of living." In fact, "no viable development is possible without sovereignty" (Farsakh, 2016, p. 59; cf. Roy, 1995). The limitation to this approach is that some level of development might facilitate sovereignty. This could conceivably occur because economic or social development could promote community integration and political capacities, with one kind of capacity facilitating the development of another (cf. Bourdieu, 1986). Dajani (2004) argues that the political weakness of Palestinians has compromised their capacity to negotiate effectively during peace deals.

Independently of the framework used to interpret the Palestinian Question, many people agree that occupation, underdevelopment, Palestinian authority corruption, and antagonisms between various Israeli and Palestinian factions are potential barriers that constrain economic development (Khan et al., 2004). Some say development contribute to limiting terrorism, conflict or at least consolidating the power of Palestinians in their dealings with Israelis (Kragin & Chalk, 2003). While others reject the ability of economics to address extremism (Piazza, 2006), development could be promoted despite or in the face of barriers: "a participatory process of development policymaking could place the war-torn economy on a footing that would allow it to cope with the constraints imposed by long-term occupation" (Khalidi & Samour, 2011, p. 18).

The current economic status of Palestine as a region under Israeli military occupation and dominated by what some consider unequal exchanges with Israel creates many challenges to higher education. The conflict and occupation have been associated with "closures and curfews that lasted for months and even years, Israeli military attacks on universities, and the arrest of teachers and students." The occupation has limited the ability of higher education in Palestine to promote economic development, with the military occupation increasingly separating "one Palestinian from the next." The Israeli government has linked "severe movement restrictions" to "security concerns." In addition, Palestinian universities have faced financial shortages. Nevertheless, universities have contributed to the development of "human capital…and maintaining people's dignity and national identity" (Alfoqahaa, 2015, 25–27, 29–30; see also Shibib, 2021). The Israel military places extensive controls on travel and imports making "so researchers find it difficult — and sometimes impossible —to get to international conferences or labs and to access research materials." A 2018 report noted that "Palestinian authorities" had "devoted little financial support to building up science in the territories" (Abbott, 2018).

3.3. Diasporic Linkages

Science diplomacy has encompassed countries' attempts "to reach out to scientists in their national diaspora." One U.S. government and scientist effort involved creation of "the Networks of Diasporas in Engineering and Science," whose goal was to boost "economic growth across a wide range of countries," linking "countries of origin and destination for members of a given diaspora" (Turekian et al., 2014). In Israel various international actors, "seeking to promote Jewish culture and the Jewish national project" helped to create the Hebrew University: "In the absence of a State, it was heavily shaped by Diaspora including its funding, governance, academic staff and students" (Bamberger, 2020, p. 79). More recently, immigrant engineers from countries like Israel and Taiwan have helped develop their home nations through work experiences in Silicon Valley in the U.S., a process called "brain circulation" (Saxenian, 2002). Diasporic relationships have helped build university ties with industry in Palestine (Sharabati-Shahin & Thiruchelvam, 2013).

3.4. University Exchanges and Organizational Commitments

University programs can be developed by transferring knowledge from one region (or university) to another. For example, in Italian universities during the period of early modern Europe (1500–1800), there was a "high proportion of foreign students—and also, in part, of professors—at the leading universities, sometimes representing half of all matriculations" (Hammerstein, 1996, p. 148). Alvin Johnson was an economist and journalist who helped establish The New School for Social Research in New York City after World War I. He also created the "University in Exile" within that institution. He did so by "offering lifesaving visas and academic positions for nearly two hundred refugees." The university acted as a solidarity system by helping provide employment or a research foundation for immigrants rendered marginalized by actions of a state (Friedlander, 2019, pp. ix–x). So, the university's *organizational commitment* was linked to a philosophical or moral principle. Organizational commitments can be "contextual and shaped by dynamics specific to certain cultural and social contexts" as well as cost reduction (Lämsä & Savolainen, 2000, pp. 300, 303). Finally, university development can be systematically advanced through aid programs at the national (Swedish) or supra-national (EU) level which compensate for market failures vis-à-vis research support (Geuna, 2001). Here the key exchange is between the state and university, potentially supporting social objectives.

3.5. Science Diplomacy

The diasporic linkages and university exchanges just described can be regarded as a form of science diplomacy. Science diplomacy can involve exchanges between nations in areas of common interest, such as "threats to the environment, human health or the safety of people" (Ruffini, 2017, p. 13). Similarly, science diplomacy can be tied to development assistance involving scientific dimensions or "enhancement of scientific literacy and capacity through the promotion of science, technology, engineering, and mathematics (STEM) education" or development to aid public health (Turekian et al., 2014). The concept gained a great deal of currency during a 2009 meeting sponsored by the Royal Society in London and the American Association for the Advancement of Science (AAAS). This meeting led to a widespread taxonomy for science diplomacy based on three core concepts: (a) "science in diplomacy," which scientists provide advice that informs and supports foreign policy objectives; (b) "diplomacy for science," where diplomacy facilitates international scientific cooperation; and (c) "science for diplomacy," where scientific cooperation improves international relations (Turekian et al., 2014).[1]

Definition (a) implies a state-led science diplomacy but leaves open the possibility that other actors might promote a state's objectives without the state initiating the action. Although a "contested" term by some, *citizen diplomacy* is "a metaphor to describe people who participate in cross-border interactions (citizen-led citizen diplomacy)" and is "a term used when private citizens are involved in state-sanctioned diplomatic interactions (state-led citizen diplomacy)." Academia is one of the areas involving such "people-to-people contact across borders" (Tyler & Beyerinck, 2016, p. 522; cf. Rana, 2011, p. 206). Thus, when academics as citizen diplomats promote international scientific cooperation, they fulfill the requirements for definition (b). Definition (c) was partially inspired by the idea of using science "to reduce tensions between Western and Muslim-majority countries, especially in the wake of the 9/11 attacks and the highly publicized war on terror, with a significant June 4, 2009, speech by President Obama focusing on science and technology servicing these objectives (Turekian et al., 2014). Therefore, if an effort builds on scientific exchanges between Western and Muslim areas, one could say it fulfils one criterion for science diplomacy. In sum, science diplomacy involves building bridges and collaborations across international boundaries and linkages among societies, with special importance placed on areas where official mechanisms are lacking (Ruffini, 2017, pp. 11–12).

The aforementioned definitions center on trans-national relations, building bridges across boundaries, mechanisms beyond official

---

1 Later elaborations included: (d) "actions designed to directly advance a country's national needs," (e) actions designed to address cross-border interests; and (f) "actions primarily designed to meet global needs and challenges" (Turekian et al., 2014).

engagements, and scientific collaborations related to shared problems and partnerships. There is no suggestion here that diplomats and the nation state are necessarily involved. In contrast, other definitions argue that "science diplomacy is a responsibility of sovereign states and most often cannot be considered on a different level than their own." By this definition, particular entities like universities cannot engage in science diplomacy, "there is a French, an American or a Russian science diplomacy, but there is not as such a science diplomacy of the *Centre National de la Recherche Scientifique*, the National Science Foundation or the Russian Academy of Sciences." Moreover, "science diplomacy should not be confounded with the multifaceted field of internationalization of research" including "international cooperation" (Ruffini, 2017, p. 15). Ruffini cites Copeland (2011) who argues that "by virtue of its direct relationship to government interests and objectives, science diplomacy differs" from such cooperation "which is sometimes commercially oriented and often without direct state participation" (Copeland as quoted in Ruffini, 2017, p. 16).

Even advocates of the state-centric view of science diplomacy acknowledge that "the international reach of a country's research opens up the best opportunities to cooperate with others" and "facilitates the access of its researchers to international scientific expertise." One argument is that "the state bears the ultimate responsibility of the national science diplomacy," but that such diplomacy can involve non-state actors, i.e. "science diplomacy must be 'integrative' in the sense that a large variety of actors must find their place in it" (Ruffini, 2017, pp. 15–16).

It can be argued that science diplomacy involves networks that link state and non-state actors. Those focusing on state-led science policy concede that: "Science diplomacy is part of soft power, and so are the international promotion of national language or cinema, etc. Thus, culture and science can be combined in a single analysis (the external action of the state) all the more so when at the operational level they use the same kind of media and tools (embassies, cooperation budgets)" (Ruffini, 2017, p. 15). A university exchange initiative could benefit from a relation to an embassy, or a state-financed cooperation budget, yet be part of a set of relationships which are not immediately initiated or organized by the state, i.e. *state inputs* can be involved. Furthermore, universities like films can involve various "producers" who generate soft power for a state, without the *state's direction*.

Some argue such science diplomacy can simply take the shape of an organizational "platform" that builds trust, "reciprocity, and an understanding and cooperativeness in which capacity-building and cross-boundary collaboration flourish" (Aukes et al., 2021, p. 12). Relevant actors can include research managers, science and knowledge institutions, scientists, and others whose daily working lives cross national borders (Aukes et al., 2021, p. 12, 2022, p. 3).

4. Methodology

This study is based on interviews and supporting secondary material. I use an oral historical approach, a method which is useful for studying various inter-organizational relationships (Perks, 2010). I interviewed the principal informant, Dr. Adi Anani, twice: once face-to-face and then in a Zoom meeting. I also conducted two additional interviews with his students.[2] Based on the theoretical framework and grounded in my empirical analysis, I created a matrix of indicators designed to ground linkages between theory and practice (Table 1). I used the inductive method of code (or indicator) development (Boyatzis, 1998).

Boyatzis, Richard E. (1998). *Thematic Analysis and Code Development*, Thousand Oaks, California: Sage.

5. Dr. Adi Anani's Science Diplomacy Efforts

5.1. Background

Dr. Adi Anani was born in Jaffa. He received his B.Sc. in engineering Alexandria University in Egypt in 1966. He came to Sweden later primarily to be a doctoral student. He gained a PhD at Chalmers University of Technology in Gothenburg in 1979 where he focused on applied electronics directed to medical electronics and biomedical engineering. He did work related to his doctoral studies in a hospital and thereafter worked in academia from 1980 to 2011. He was a visiting professor at the University of Wisconsin-Madison. Between his doctoral studies and his final academic position, he worked at universities in the Middle

2 This material is not used here for reasons of space, but confirmed the larger points made by Anani.

Table 1: Methodological System

Key Theory	Key Factor	Key Indicators
Palestinian development and peace/diplomacy	Trust building	Statements indicating a trusting relationship was established
	Corruption	Statements about actions taken to address or sidestep influence of corrupt authorities
	Border curfews, closures, movement delays, security concerns	Statements about actions taken to address or sidestep these factors
	Resource deficits in peripheral regions	Discussions of actions taken to compensate for such deficits
Diasporic linkages	Solidarity or identity system	Diasporic commitments revealed in narratives to explain actions
University exchanges	University commitments tied to cost reduction or a moral context; Exchange based on information or financial transfer	Statements linking engagement to cost reduction or a moral context. Migration of researchers from one region to another such that knowledge is acquired or exchanged.

Source: Author based in part on Boyatzis (1998) and material cited in theory section above.

East including Yarmouk University in Jordan and Kuwait University in Kuwait. From 1990 to 2011 he worked at Umeå University in the Department of Applied Physics and Electronics. He was a guest professor at University of Wisconsin Madison, USA, 1993 and Bedford University, U.K. in 1994 (Anani, 2019, 2022).

5.2. Diasporic Commitments

The journal *Higher Education* published an article in 2013 identifying Anani as a central member of the Palestinian diaspora in Sweden who had "initiated some efforts to establish…connections with institutions inside Palestine to help in capacity building and human resource development" (Sharabati-Shahin & Thiruchelvam, 2013, p. 625). He has been "probably one of" the few Palestinian academics based in Sweden who has worked "for cooperation with Palestine." He "did not meet" many others working to support technological development in Palestine. He argues that there is no real formal network to encourage economic exchanges between Palestinians living in Sweden and persons in Palestine. One reflection of this absence is that Anani has not been contacted by the Palestinian Embassy, aid agencies, or politicians about the projects he has been involved in, although he is not generally familiar with current Swedish engagements in this area. Anani argues "Palestinian universities are the core of…development in Palestine" (Anani, 2019).

Anani was engaged in three principal solidarity and cooperation projects with Palestine. These were motivated by his desire to help Palestine because he "originally came from Palestine." He "tried to help," with this motivation most focused on the Polytechnic in Hebron (Hebron Polytechnic University or Palestine Polytechnic University, PPU). PPU was "principally disconnected from the world" and "had no connection whatsoever." As he explained, "so I tried to have some connection with them so that they [could] feel better," including their "spiritual life," and not "feel isolated and so on." He also wanted to gain the benefits from a mutual learning experience linking Swedes and Palestinians. At the time of his engagement, he did not rely on any Palestinian office in Sweden and made all his contacts "personally." His personal commitment overlapped with a professional engagement and were part of his "academic work, more or less" (Anani, 2019, 2022).

5.3. University Exchanges: Umeå University and the SIDA Project

From the Palme era and after the Six-Day War (1967), Sweden became more engaged in Palestinian causes (Badarin, 2020). Umeå University was motivated to engage in projects of the national aid agency SIDA because they wanted researchers "to apply for work with the Third World." The Palestine project was part of a group of such projects, with one or two others in Latin America. Umeå University received money from SIDA which funded the university to identify the ability "to make projects for less costs" than they were accustomed to spending. SIDA's goal in financing the PPU biomedical laboratory was "to make projects with the least possible amount of money." Anani "managed to do that" as he received only 300,000 SEK for "the project to make exchanges and to start a laboratory" which SIDA thought was "impossible." He therefore revealed that such projects could be made less expensively than was considered (Anani, 2022).

During the 1970s, modern universities emerged in Palestine. One key area of success in Palestinian higher education has been the development of multiple medical schools, where "trauma medicine is recognized in Palestine as the best in the world" (Alfoqahaa, 2015, pp. 27–28). Anani's first project directly involved SIDA and began during the 1995–1996 period and lasted three years. The goal of this pilot project began by equipping a laboratory to promote biomedical engineering at Hebron Polytechnic. Anani's work "was unpaid" by this project, however, i.e. he covered his expenses through his regular university job and "there was no direct payment" for him to engage in the project (Anani, 2019). Biomedical engineers invent, design, and build new technologies for diagnosis, treatment, and study of human disease." These persons usually "work as part of a team of engineers, scientists, and physicians," with the engineer in such teams "responsible for converting new knowledge into a useful form" (Saltzman, 2009, p. 6).

Anani selected PPU because normally Birzeit University is "the first" university that outsiders choose to collaborate with. Birzeit University even visited Anani and asked him "to cooperate with them, but [he] thought they had it easier" in their ability "to establish contact with universities in the United States and Europe." In contrast, Hebron Polytechnic University (PPU) "was a little bit isolated, on the southern part of the West Bank and they did not have as good contact with the universities on the outside." Anani believed that contacts with the outside would give Palestinians a broader outlook and enabled them "to see how it works in other countries, how they can cooperate" (Anani, 2019).

The instrumentation lab in Hebron helped engineering students understand the principles of electrocardiogram (ECG), electroencephalogram (EEG), Electromyography (EMG) and the design of various components. The laboratory provided knowledge to students about the basics of biomedical engineering, such as "how to take care of instruments" at medical facilities. Students learned both how to use diverse kinds of medical equipment and how such equipment was built. Palestinian students learn about biomedical equipment design in universities, but there are no companies producing such equipment in Palestine, "so there's no industry where they can use their achieved design qualifications." Equipment is imported for the most part from foreign companies in Germany, the United States, Japan and elsewhere. Yet, training in the laboratory has helped Palestinian engineers understand how to use "new equipment" and how to take care of it, primarily at hospitals. Students learned about the design of components, e.g. training was provided "in how to design" and "start building amplifiers used in biomedical engineering" and "how to filter some noise coming from" equipment, etc. (Anani, 2022).

After the project was completed, SIDA was re-organized which led the persons who Anani worked with to change their positions. As a result, "there was no follow up" regarding the pilot project on SIDA's side. Anani ceased having direct contact with SIDA except when he learned that one could gain money from the Linnaeus-Palme program through its financier SIDA. This inquiry led to his next solidarity initiative (Anani, 2019).

5.4. University Exchanges: The Linnaeus-Palme Project

The Linnaeus-Palme initiative is "an exchange programme in which Swedish universities cooperate with universities in developing countries" (Studera.nu, 2019). These projects aimed at universities in the Global South were administered by the *Universitets- och högskolerådet*, i.e. the Swedish Council for Higher Education. This program financed Anani's travel to Palestine. Anani's project involved an exchange of students and teachers in Sweden and Palestine focusing again on PPU. Three teachers from Hebron were involved in this project starting around 2003. Anani had tried to finance two Umeå University students through the SIDA "minor field studies" program that allowed students to do their graduation project overseas. After "a bomb exploded in Jerusalem…for security reasons" they could not participate. Four or five Palestinian students ended up coming to Sweden, however. Four of these managed to gain a master's degree as a result of their engagement even though the Linnaeus-Palme program does not normally generate master's graduates. Rather, "it is an exchange for a maximum of two terms." Anani explains, "the students were very eager to make something out of this—not just come and go." As a result, he "helped them being registered as students and they managed to leave the university with their Masters" degree. Two of these students went back to Palestine and two remained working in Sweden. These students mostly focused on interaction and design as well as computer programming. One of the returnees to Palestine

(Abdelfattah Najjar) still works presently at PPU in media-related activities. Anani initiated and had become involved in Umeå University's media engineering program and thereafter in a master's degree in media applications. The student returned, adapted the courses he studied in Sweden to Palestine and is presently a successful lecturer at the Department of Multimedia and Graphic Design, PPU (Anani, 2019, 2022).

While the training of Palestinian students in Sweden was a successful format for an economic solidarity system, "the scale was too small." There were only about three to four university lecturers from Umeå University involved. One of the Palestinian participants, Alaa Halawani, was a post-doctoral researcher at Umeå University for five years, and then worked for a period of time within Swedish industry before returning to PPU. His trajectory shows how Sweden and Palestine both gained from the initiative, i.e. leading to mutual benefits in research. Halawani, therefore, had what amounted to a kind of internship (Anani, 2019).

5.5. University Exchanges and Negotiating Security: The Tempus Project

The third project Anani led involved the Tempus project, ICT Learning Pentagram, ICT-LEAP, at the European Union. The EU Directorate-General for Education and Culture awarded the project SEK 3.6 million in January 2006. This project was a consortium that linked Umeå University, Dortmund University in Germany, Jordan University of Science and Technology (JUST), Irbid in Jordan and PPU and Al-Quds University in Palestine. The project concerned media technology and bio-media including biometrics. The goal was to establish online teaching and create so-called "E-learning centers." The project received a write up in the *Västerbottens-Kuriren* and other technical journals (Anani, 2019; Dahlberg, 2009; Karlsson et al., 2009).

This project led to the development of a course given simultaneously at these diverse universities using video conferencing. Video conferencing equipment was installed at PPU and Al-Quds University. The bio-media course, which was the project's main product, involved the study of the signals that can be understood by human beings (information reception) and the natural or physiological signals of human beings like fingerprints and registering of the iris (Anani, 2019). The Tempus project and related research illustrated how to overcome communication barriers when there was "no infrastructure for Internet and mobile communication or where the cost to use the existing infrastructure is too high." The solution was "based on wireless ad-hoc networks" (Karlsson et al., 2009, p. 22).

5.6. Outcomes Related to a Science Diplomacy Framework

The Hebron program triggered by Anani's project still exists to this day: "The BME Program at PPU was founded in 1997 to maintain pace with global developments in the fields of technology pertaining to health-care and biomedical equipment. The first of its kind in Palestine, the BME Program leads to a B.Sc. degree in Biomedical Engineering. The curriculum has been prepared to fit the needs of the local medical community; therefore, it focuses on medical instruments and their uses in addition to development and maintenance" (Keystone Healthcare Studies, 2017).[3] One of the key supporters, Dr. Daoud Zatari, later became the President of the University and Mayor of Hebron. During this project three PPU teachers took part in an exchange activity with Umeå University, the activity was financed by the project (Anani, 2019). Anani believed his PPU project could facilitate conditions for peace "because people felt themselves [to be] respected," they had freedom of movement and could meet other people and cooperate with European universities. This allowed participants "to feel the meaning of peace because they had no peace at all, especially at that time and not even now." The project only "indirectly" contributed to peace because being limited to academic realities, it could not "affect the political situation" (Anani, 2022).

Anani's leveraged his power as a Swedish citizen to leverage Swedish state capacities to negotiate a solution to problems of war-based occupation. In the 2009 inauguration of the Tempus video project, students, teachers and principals at five universities gathered as part of a video conference network. This inauguration was a year late because of Israeli authorities delayed equipment for a year. This meant that Palestinian universities in the West Bank and in Jerusalem could not gain access to video conferencing equipment. At the time, Anani explained how Sweden helped this project negotiate occupation obstacles: "After, among other things, Swedish pressure, the Israelis let the equipment through, except for the wireless microphones that were not allowed in for security reasons" (Dahlberg, 2009). Anani's intervention with the Swedish

---

3 According to this website, this program appears to be the only such biomedical program in Palestine (Keystone Healthcare Studies, 2022).

embassy led the Israeli customs office to release the equipment. Hebron Polytechnic's president issued "a declaration" that the equipment would "not be used for terrorism and so on." As a Swedish citizen, Anani could gain the embassy's help. The university's leaders also sidestepped problems by buying biomedical equipment from an Israeli company. Anani interceded with the Swedish embassy in Tel Aviv to help Hebron gain access to the outer world. He helped Hebron students get visas through Swedish officials and even booked tickets from travel agencies through SAS's Jerusalem office (Anani, 2022).

The Tempus project itself was partially designed to overcome occupation limitations as Anani explained: "Many times [students] find it difficult to get to the university because they do not get through the Israeli roadblocks. Now they can see lessons they missed." While two of the participating universities were only 3.5 km apart, it could take almost a day to travel between because of closed roads and roadblocks, if you could even reach your destination (Dahlberg, 2009; cf. Anani, 2019).

5.7. Problems of Palestinian Development

Anani was able to sidestep problems of development centered on corruption by designing networks that tried to bypass authorities:

> I tried to do the project without interference from any officials [and pursued] the project [so it could be organized] without any corruption. That's why I made all the contacts personally, with persons *directly*, so that that the money and the exchange and everything were done from here, only from Sweden...And it did not go through any other channel, the money went directly to the guys involved.

By directly making his own contacts, Anani could bypass corruption caused by using government filters to distribute or use funds destined for Palestine or Palestinian project participants. He also addressed corruption by making ethical decisions. One potential problem which could be perceived as a form of corruption emerged: A person "who was not part of the academic staff but was a high-ranking person at the Polytechnic... wanted to visit Sweden on the project list and I refused explaining that this was not in line with the exchange program goals and regulations" (Anani, 2022).

6. Conclusions: University Exchanges as Science Diplomacy

My first research question was whether university-based scientific exchanges contribute to science diplomacy. I argue that they do so for several reasons. First, Anani's efforts addressed threats to human health and contributed to public health, human security and safety. He engaged in a kind of networking taking the form of the aforementioned "science in diplomacy" which contributed to Sweden's goals of solidarity with the Global South as manifested in the actions of universities (Umeå University) and aid organizations (SIDA). Anani was engaged in "citizen diplomacy" involving person-to-person contacts across borders. He also engaged in "science for diplomacy" to the extent this his scientific exchanges built up cooperation, built relationships and partnerships between Western (i.e. Swedish) and more Muslim (i.e. Palestinian) regions, with potential dividends for Sweden's long-term commitment to Palestinian development (cf. Badarin, 2020). Anani's experience illustrates how he was part of a less formalized foreign policy agent (the universities which are state organized by not usually considered "political actors") but leveraged more formalized actors in foreign policy (with help from SIDA and the Swedish Embassy in Israel). It seems safe to say such cooperative projects helped build trust between Sweden and Palestine, contributing to Swedish "soft power" or reputational capital.

I have addressed both research questions in the empirical analysis of Anani's efforts which are summarized in Table 2. These show that his work advanced trust building, sidestepped corruption, negotiated Israeli security concerns and occupation, and contributed to Palestinian development with limited resources. Both diasporic and university linkages contributed to these goals. Anani's trajectory clearly shows how European science diplomacy could benefit from cultivating the action of various diasporic and university entrepreneurs who build solidaristic, development efforts across borders. Anani's experiences also show how such informal efforts often are ignored and underappreciated as states may focus more on formalized approaches that don't always prove successful.

Table 2: Key Empirical Findings on Anani's Science Diplomatic Efforts

Key Theory	Key Factor	Empirical Data
Palestinian development, peace/ diplomacy	Trust building	Experiences for Palestinians which facilitated freedom of movement, travel outside Palestine and cooperation with others (Swedes and other Europeans).
	Corruption	One project sidestepped this problem through Anani's oversight and decision to make arrangements by himself without heavy reliance on Palestinian authorities.
	Border curfews, closures, movement delays, security concerns	These factors associated with occupation constrained the project but were addressed by creating a diplomatic network building on Anani's status as a Swedish citizen, university linkages, and dealmaking with Israeli entities. By advancing video conferencing some of the ways movement restrictions limited learning could be overcome.
	Resource deficits in peripheral regions	The biomedical project contributed to a laboratory and educational program where none existed before in Palestine. The Tempus project showed how Palestine could leverage education inputs from other European regions through video conferencing.
Diasporic linkages	Solidarity or identity system	Adani leveraged Swedish resources to support development in Palestine, based in part of a shared background as well as larger humanitarian and national concerns.
University exchanges	University commitments tied to cost reduction or a moral context;	Adani became engaged in Palestine given his background, concern for Palestinians and as part of attempts by various parties to promote low-cost projects.
	Exchange based on information or financial transfer	Adani organized exchanges between Umeå and PPU which created the first biomedical laboratory in Palestine. Students learned about equipment use and design. The Linnaeus-Palme project provided direct training for Palestinian students.

References

Abbott, Alison. (2018). In the Palestinian territories, science struggles against all odds. *Nature.* November 14. Accessible at: https://www.nature.com/articles/d41586-018-07350-9. Accessed 11 June 2022.

Alfoqahaa, Sam Abd Al-Qadir A. (2015). Economics of higher education under occupation: The case of Palestine. *Journal of Arts and Humanities*, 4(10), 25–43.

Anani, Adi. (2019). Interview with author, Solna, Sweden, February 18.

Anani, Adi. (2022). Zoom meeting with author, February 8.

Aukes, Ewert, Stefan, Kuhlmann, & Gonzalo, Ordóñez-Matamoros. (2021). Reshape science diplomacy. *Research Europe*, July 22.

Aukes, Ewert, Wilsdon, James, Ordóñez-Matamoros, Gonzalo, & Kuhlmann, Stefan. (2022). Global resilience through knowledge -based cooperation: A new protocol for science diplomacy. *F1000 Research*, Last updated, February 10. https://doi.org/10.12688/f1000research.55199.1

Badarin, Emile. (2020). States Recognition in Foreign Policy: The Case of Sweden's Recognition of Palestine. *Foreign Policy Analysis*, 16(1), 78–97.

Bamberger, Annette. (2020). *Diaspora, state and university: An analysis of internationalisation of higher education in Israel.* (PhD Thesis). London: University College London Institute of Education, June 1.

Bourdieu, Pierre. (1986). The forms of capital. In John G. Richardson (ed.), *Handbook of theory and research for the sociology of education* (pp. 46–58). New York: Greenwood.

Boyatzis, Richard E. (1998). *Thematic analysis and code development*. Thousand Oaks, California: Sage.

Copeland, Daryl. (2011). *Science diplomacy: What's it all about?* CEPI-CIPS policy brief 13. http://cips.uottawa.ca/wp-content/uploads/2011/Copeland-Policy-Brief-Nov.11.5.pdf.

Dahlberg, Carina. (2009). *Direktsänd invigning av E-lärcentrum från Umeå, Västbanken, Jerusalem, Jordanien och Tyskland*. Umeå University, May 11. Accessible at: https://www.umu.se/nyheter/direktsand-invigning-av-e-larcentrum-fran-umea-vastbanken-jerusalem-jordanien-och-tyskland_5837244/. Accessed 14 June 2022.

Dajani, Omar M. (2004). Understanding barriers to peace: A Palestinian response. *Negotiation Journal*, 20(3), 401–408.

Farsakh, Leila. (2016). Palestinian economic development: Paradigm shifts since the first Intifada. *Journal of Palestine Studies*, 45(2), 55–71.

Friedlander, Judith. (2019). *A light in dark times: The new school for social research and its university in exile*. New York: Columbia University Press.

Geuna, Aldo. (2001). The changing rationale for European university research funding: Are there negative unintended consequences?. *Journal of Economic Issues*, 35(3), 607–632.

Hafez, Mohammed M. (2006). *Manufacturing human bombs: The making of palestinian suicide bombers*. Washington, D.C.: United States Institute of Peace Press.

Hammerstein, Notker. (1996). Relations with authority, Chapter 3. In *A history of the university in Europe, volume ii universities in early modern Europe* (1500–1800) (114–153). Cambridge: Cambridge University Press.

Hsu, Henda Y. & David McDowall. 2020. Examining the state repression-terrorism nexus: Dynamic relationships among repressive counterterrorism actions, terrorist targets, and deadly terrorist violence in Israel. *Criminology and Public Policy*, 19(2), 483–451.

Israeli, Raphael. (2002). A manual of Islamic fundamentalist terrorism. *Terrorism and Political Violence*, 14(4), 23–40.

Karlsson, Johannes, Anani, Adi & Li, Haibo. (2009). Enabling real-time video services over ad-hoc networks opens the gates for E-learning in areas lacking infrastructure. *International Journal of Interactive Mobile Technologies (iJIM)*, 3(4), 17–23.

Keystone Healthcare Studies. (2017). *Bachelor in Biomedical Engineering (BME)*. Keystone Education Group, Fornebu, Norway, December. Accessible at: https://www.healthcarestudies.com/Bachelor-in-Biomedical-Engineering-(BME)/Palestinian-Territories/Palestine-Polytechnic-University/. Accessed 16 June 2022.

Keystone Healthcare Studies. (2022). *Bachelor in Biomedical Engineering in Palestinian Territories*. Fornebu, Norway: Keystone Education Group. Accessible at: https://www.healthcarestudies.com/Bachelor/Biomedical-Engineering/Palestinian-Territories/. Accessed 18 February 2019 and 16 June 2022.

Khalidi, Raja & Samour, Sobhi. (2011). Neoliberalism as liberation: The statehood program and the remaking of the Palestinian National Movement. *Journal of Palestine Studies*, 40(2), 6–25.

Khan, Mushtaq Husain with George Giacaman and Inge Amundsen (eds.) (2004). *State formation in Palestine: Viability and governance during a social transformation*. New York and London: Routledge Curzon.

Kragin, Kim & Chalk, Peter. (2003). *Terrorism and development: Using social and economic development to inhibit and resurgence of terrorism*. Santa Monica, CA: Rand Corporation.

Lämsä, Anna-Maia & Savolainen, Talna. (2000). The nature of managerial commitment to strategic change. *Leadership & Organization Development Journal*, 21(6), 297–306.

Lloyd, David. (2012). Settler colonialism and the state of exception: The example of Palestine/Israel. *Settler Colonial Studies*, 2(1), 59–80.

Pape, Robert A. (2006). *Dying to win: The strategic logic of suicide terrorism*. New York: Random House Trade Paperback.

Perks, Bob. (2010). 'Corporations Are People Too!': Business and corporate oral history in Britain. *Oral History*, 38(1), 36–54.

Piazza, James A. (2006). Rooted in Poverty?: Terrorism, Poor Economic Development, and Social Cleavages. *Terrorism and Political Violence*, 18(1), 159–177.

Rana, Kishan S. (2011). *21st Century Diplomacy*. London: The Continuum International Publishing Group.

Roy, Sara. (1995). *The Gaza Strip: The political economy of de-development*. Washington, D.C.: Institute for Policy Studies.

Ruffini, Pierre-Bruno. (2017). *Science and diplomacy: A new dimension of international relations*. Cham, Switzerland: Springer International.

Saltzman, W. Mark (2009). *Biomedical engineering: Bridging medicine and technology*. Cambridge: Cambridge University Press.

Saxenian, AnnaLee. (2002). Brain circulation: How high-skill immigration makes everyone better off. *The Brookings Review*, 20(1), 28–31.

Sharabati-Shahin, Mervat H. N. & Thiruchelvam, K. (2013). The role of Diaspora in university-industry relationships in globalised knowledge economy: The case of Palestine. *Higher Education*, 65(5), 613–629.

Shibib, Khalid. (2021). From higher education in historic Palestine towards a Pan-Palestinian higher education. *Contemporary Arab Affairs*, 14(2), 21–54.

Studera.nu. (2019). *Linnaeus-Palme.* Accessible at: https://www.studera.nu/startpage/study-abroad/exchange-programmes/linnaeus-palme/. Accessed 18 February 2019.

Turekian, Vaughan C., Gluckman, Peter D., Kishi, Teruo, & Grimes, Robert W. (2017). Science diplomacy: A pragmatic perspective from the inside. *Science and Diplomacy*, 6(4), [online]. Available from: http://www.sciencediplomacy.org/article/2018/pragmatic-perspective. Accessed 19 February 2022.

Tyler, Melissa Conley & Beyerinck, Craig (2016). Citizen diplomacy. In Costas M. Constantinou, Pauline Kerr & Paul Sharp (Eds.), *The Sage handbook of diplomacy* (pp. 521–529). Los Angeles: Sage Publications.

Chapter 9

Parliamentary Diplomacy and Personal Data Protection in the Digital Age: An Unexplored Opportunity for Upholding Human Rights in the European Union and Beyond?

Mónica Velasco-Pufleau

Abstract: The chapter explores the role of the European Parliament in upholding the right to privacy and personal data protection in the EU and its relations with the wider world, by tracing dialogue and cooperation in this realm at the interparliamentary level in practice. Drawing from the work and outcomes of the Euro-Latin American Parliamentary Assembly over the last decade (2013–2023), the chapter shows how parliaments can play a relevant role in promoting these rights in the digital age not only domestically, but also internationally through parliamentary diplomacy. By doing so, the chapter advances the existing literature on this field, introducing a fined-grained examination of a topic that has been largely overlooked. This holds particularly true concerning the studies addressing parliamentary diplomatic activity in the framework of the EU-Latin America and the Caribbean relations.

Keywords: European Union, EuroLat, parliamentary diplomacy, privacy and personal data protection, EU-LAC Digital Alliance.

1. Introduction

Unlike other global players, the European Union (EU) strongly stands as a normative actor in international relations, promoting principles and values such as respect for human rights –also referred to as fundamental rights in the EU's legal order and policy– (Manners, 2008).[1] This approach has been incorporated in the EU Treaties themselves,[2]

---

1 On an in-depth analysis of the dual rights language used by the EU, see Wouters and Ovádek (2021, pp. 70–77).

2 OJ C 202/1, 7.6.2016 (consolidated versions 2016).

which provide that the Union shall promote its values in its relations with the wider world, as well as contribute to the protection of human rights, among others (Articles 2 and 3(5) Treaty on EU –TEU–). In the same vein, it is envisaged that the "Union's action on the international scene shall be guided by the principles which have inspired its own creation, development and enlargement, and which it seeks to advance in the wider world", including "the universality and indivisibility of human rights" (Article 21(1) TEU). Within that context, it "shall define and pursue common policies and actions and shall work for a high degree of cooperation in all fields of international relations, in order to […] consolidate and support democracy, the rule of law, *human rights* and the principles of international law" (Article 21(2)(b) TEU, emphasis added). The principles and objectives contained in Article 21 TEU reach both the EU's external action and the external aspects of its internal policies (Article 21(3) TEU, see also Article 205 Treaty on the Functioning of the EU –TFEU–).

In particular, the right to personal data protection is a fundamental right enshrined in Article 8 of the Charter of Fundamental Rights of the EU.[3] The Charter is the "centrepiece of EU fundamental rights law and policy", which acquired a legally binding status and the same legal value as the EU Treaties since the entry into force of the Lisbon Treaty in 2009 (Wouters & Ovádek, 2021, p. 100). Moreover, Article 7 of the Charter provides for a right to privacy, encompassing private and family life, home and communications. Together, these two articles grant a legal basis for the EU's data protection regime (Mildebrath, 2023, p. 1).

Internationally, the right to privacy is part of major human rights instruments, such as the 1948 Universal Declaration of Human Rights proclaimed by the United Nations General Assembly. Similar provisions have also been included, for example, in regional human rights treaties. The latter includes the 1950 European Convention for the Protection of Human Rights and Fundamental Freedoms, which serves as "a formal inspiration" of the general principles of fundamental rights of EU law (Bartels, 2014, pp. 1075, 1077).

This chapter explores the role of the European Parliament (EP) in upholding the right to privacy and personal data protection in the EU and its relations with the wider world, by tracing dialogue and cooperation in

[3] OJ C 202/389, 7.6.2016.

this realm at the interparliamentary level in practice. Drawing from the work and outcomes of the Euro-Latin American Parliamentary Assembly (EuroLat) over the last decade (2013–2023), the chapter shows how parliaments can play a relevant role in promoting these rights in the digital age not only domestically, but also internationally through parliamentary diplomacy –hereafter PD–. By doing so, the chapter also advances the existing literature on PD, introducing a fined-grained examination of a topic that has been largely overlooked at the time of writing. This holds particularly true concerning the studies addressing parliamentary diplomatic activity in the framework of the EU-Latin America and the Caribbean (LAC) relations (see, e.g., Luciano, 2017; Müller, 2019).

The research is qualitative and is widely based on the analysis of primary data from the EU institutions, excepting the second section. Mainly, these EU institutions include the European Commission (EC), acting individually or jointly with the High Representative of the Union for Foreign Affairs and Security Policy (HR), and the EP in what relates to the EuroLat.

The work is divided into five sections. After this introduction, the second section presents the concept of PD. Special attention is paid to the EP, acknowledged as a world leader in the field (Jančić, 2017). The section also refers to the EuroLat, whose creation in 2006 was strongly advocated by the EP to provide the EU-LAC Strategic Partnership with a parliamentary dimension (Fernández Fernández, 2022). The third section addresses, in a nutshell, the inter-governmental digital agenda between the two regions, focusing on the latest developments at the time of writing. Notably, developments include the EU-LAC Digital Alliance endorsed by the third Summit between the Heads of State and Government of the EU and the Community of Latin American and Caribbean States (CELAC, by its Spanish acronym) held in Brussels in July 2023. The fourth section closely analyzes the work and outcomes of the EuroLat regarding the bi-regional digital agenda, especially its contributions to the protection of privacy and personal data. The concluding remarks are presented in section fifth.

2. PD and the EP: The Case of the EuroLat

There is no single definition of PD, although it is an "ever-expanding worldwide phenomenon" (Stavridis, 2021, p. 227). Only over the last decade and a half, scholars have made conscious efforts to conceptualize it, focusing on its main characteristics, functions and means (Ibidem pp. 238–239, 249). For instance, drawing from a series of empirical studies on PD in European and global governance published in early 2017, Stavridis (2017 pp. 369–374) concluded that there are four main characteristics of PD. First, flexibility, which suggests that parliamentarians have more room to operate in sensitive and difficult situations, since they are not necessarily committed to a specific government. Second, informality, which refers to the use of informal means that may be as important as formal ones in shaping international affairs. Third, the consideration of whether PD is aimed at complementing state diplomacy, or whether it should be independent from it, with supporters and detractors conferring arguments towards both approaches. Fourth, the multiplicity of levels and actors inherent to PD, including sub-national, national, regional, and international initiatives, as well as individuals and groups of individuals, state and non-state actors, to name a few (see also Stavridis 2019, 2021).

The diplomatic means employed by parliaments are also various. Yet, they can be encompassed in two main categories, according to the PD literature. On the one hand, there are those used in general in the field of foreign policy, such as the ratification of international agreements, plenary debates and resolutions, hearings, written and oral questions, reports, and budget approvals. On the other hand, there are those specific for the external action of parliaments, including missions to third countries and international organizations, friendship groups, and delegations to maintain relations with peers from individual countries or under the umbrella of international parliamentary assemblies (Stavridis, 2019, pp. 192–193). In the case of the EP, others have distinguished between the establishment of diplomatic contacts worldwide both at bilateral and multilateral levels and non-legislative instruments, namely resolutions, reports, debates, hearings and fact-finding missions, among others (Jančić, 2017, p. 23).

Finally, the functions of PD range from "political accountability for foreign and security policies" and "legitimization of (inter-)regional and global integration processes" to "contribution to conflict resolution" (Stavridis, 2017, pp. 375–381; see also Stavridis, 2021). Another important function relates to parliaments acting as "moral tribunes" in international relations, in view of which they "introduce important normative elements far from the traditional premises and prescriptions of realism while considering human rights and democratization as components of a more human and moral international system, if only at a discursive level in many cases" (Stavridis & Fernández Molina, 2013, p. 154; see also Stavridis, 2017, p. 378; 2019, pp. 200–201; 2021, p. 244).

The EP alone has 45 standing delegations at the time of writing.[4] This number signals "a dramatic growth in the importance of the external representation of the Parliament since the first direct elections' (Corbett et al., 2016, p. 207) in 1979, when there were over twenty delegations. This contrast increases even more when considering that delegations only reached over seven prior to such elections (Corbett, 1998, p. 85).

Although the most obvious task of delegations is to maintain and develop the EP's international contacts with third countries, regions, and organizations, they shall also "contribute to promoting in third countries the values on which the European Union is founded"; according to the principles governing their activities (EP Conference of Presidents 2015). These values include respect for human rights, such as the rights to privacy and personal data protection discussed above. In other words, "the delegations provide one means for the Parliament [...] to assist the EU as a whole in the pursuit of its objectives", with human rights being among their "specific areas of interest" and "a long-standing concern" (Corbett et al., 2016, pp. 206, 218).

In practice, such moral tribune role –as called by the literature– has been observed by scholars in relation to different parts of the globe, including Libya (Stavridis & Fernández Molina, 2013), Mexico (Velasco-Pufleau, 2017) and Venezuela (Lamoso González & Stavridis, 2022). In all cases, the EP sought to exert influence over the third countries' governments as well as other EU institutions with greater or lesser success. The EP's capacity to shape both the internal and external dimensions of the EU through PD activities has also been evident in the oversight of EU's international agreements, for instance (Velasco-Pufleau, 2023).

Specifically, nine EP's standing delegations exist to ensure dialogue with LAC.[5] Among them is the Delegation to the EuroLat –known as DLAT– currently composed of 75 members representing all political

---

[4] The full list is available at: https://www.europarl.europa.eu/delegations/en/list/byname (Accessed: 8 December 2023).

[5] The full list is available at: https://www.europarl.europa.eu/delegations/en/list/byregion (Accessed: 8 December 2023).

groups in the EP as well as non-attached members, thereby reflecting a high degree of political pluralism.[6] The other half of the EuroLat's membership –this is, 75 members– comes from the Latin American component, which consists of a delegation composed of representatives of the Latin American and Caribbean Parliament –previously, Latin American Parliament–, the Andean Parliament, the Parliament of the Southern Common Market, the Central American Parliament, and the Chilean and the Mexican Congresses (Fernández Fernández, 2022, pp. 379–380). The Executive Bureau is composed of one president and seven vice-presidents from each regional component (EuroLat, 2018a, Article 4(1)).

The EuroLat held its constituent meeting in Brussels in November 2006, becoming one of the various multilateral assemblies established under the auspices of the EP's external relations (see Corbett et al., 2016, pp. 211–215). It emerged from the series of biennial EU-Latin American Interparliamentary Conferences held between 1974 and 2005, characterized for being the first-ever –and for more than a decade, the only– institutionalized forum for political dialogue between both regions (Fernández Fernández, 2022, p. 376).

According to its Rules of Procedure (EuroLat, 2018a), the EuroLat is "the forum for parliamentary discussion, consultation, supervision and monitoring in respect of all questions of concern to the Bi-regional Strategic Partnership" (Article 3). To this end, it meets –in principle– at plenary level once per year, alternating between a LAC country and the EU, preferably in the country hosting the EU-CELAC Summit when applicable (Article 7(1)(2)). The plenary work is prepared by four standings committees (Article 23(1)). The Executive Bureau may also decide to set up, among others, working groups on specific topics charged with the task of drawing up reports and motions for resolutions or recommendations addressed to the Assembly (Article 25(1)). Other official EuroLat's bodies include the Euro-Latin American Women's Forum established to discuss "subjects related to aspects of life that have an impact on women's life" during the EuroLat's sessions (EuroLat, 2023a).

Moreover, the EuroLat's Rules of Procedure (EuroLat, 2018a) provide that the Assembly shall keep close ties with the institutions and bodies

---

6 The full list is available at: https://www.europarl.europa.eu/delegations/en/dlat/members (Accessed: 8 December 2023).

of the Bi-regional Strategic Partnership, including the EU-CELAC Summit (Article 5). The EuroLat can adopt non-binding resolutions and recommendations for the attention of the former institutions and bodies, as well as draft opinions and proposals at their request (Article 3, see also Articles 16(1)(5) and 22). Likewise, the Executive Bureau is entitled to adopt messages to the EU-CELAC Summit at the initiative of the Co-presidents (Article 17), who may also issue joint statements –usually, in the form of a declaration– on any matter concerning the Partnership (Article 18).

In the same vein, all the EuroLat's members have the possibility of tabling questions for written or oral answer to the ministerial bodies of the Latin American regional integration processes, the Presidency-in-Office of the Summit, the Council of the EU, and the EC (Ibidem, Articles 20 and 21). In turn, representatives of these institutions and bodies may take part in the sessions and meetings of the Assembly and its bodies (Articles 5 and 14(2)). The same applies to any other person invited by the Assembly; for instance, to provide first-hand information about situations that are of concern (Article 25(3)). It has also become common that the Co-presidents are invited to attend high-level bi-regional meetings. This happened in the EU-LAC Informal Ministerial Meeting held in Berlin in December 2020. Another relevant example is the third EU-CELAC Summit (Brussels, July 2023) that ended an eight-year period in which no summit was held, the last one taking place in June 2015. The declaration issued by the third Summit (Brussels, July 2023) recognizes PD "as an important dimension of the relationship", including "the constructive role" played by the EuroLat (Council of the EU, 2023, para. 34).

As described in the next section, a major contribution of the long-awaited third EU-CELAC Summit (Brussels, July 2023) was made precisely within the digital realm, by supporting the Digital Alliance between the EU and LAC.

3. Recent Developments in the EU-LAC Digital Agenda in a Nutshell: The Digital Alliance

Indeed, a key achievement of the third EU-CELAC Summit (Brussels, July 2023) was the endorsement of a Joint Declaration on the EU-LAC Digital Alliance by all EU Member States and 20 LAC countries (EC,

2023b), following the launching of the initiative in Bogotá in March 2023 (EC, 2023a).[7] Although previous bi-regional summits had to some extent referred to the need of strengthening cooperation regarding the digital economy and the reduction of the digital divide (see, e.g., Council of the EU, 2013a, para. 18; 2013b, section 1; EU-CELAC, 2015a, paras. 43 and 62; 2015b, section 1), the Joint Declaration gave unprecedented political support to the establishment of a first-ever digital partnership between countries of both regions in a context of recovery from the COVID-19 pandemic. This pandemic severely affected LAC, while changing trends of digitalization in society and economy globally (EC, 2021a, p. 1; 2021b, p. 8). As for the EU, it exposed vulnerabilities and high-risk dependencies regarding its digital space, triggering specific EU's targets for its digital transformation by 2030 that stressed the importance of international partnerships with like-minded countries to seize the digital future (EC, 2021a, pp. 1, 18–20; EC & HR, 2023, p. 9).

Against this backdrop, the EU-LAC Digital Alliance was conceived as a comprehensive framework for high-level dialogue and cooperation to enable inclusive digital transformation in LAC based on the "EU's value-based and human-centric approach" to digitalization, while promoting EU's objectives and strategic interests at bi-regional and international levels (EC, 2021b, p. 15). According to the EC, this EU's approach "is gaining ground in the region, fitting with the values of LAC societies and the EU-LAC partnership" (Ibidem). Consolidating the Digital Alliance is therefore crucial to address current challenges in LAC, and to secure the EU a position as key digital partner in the region vis-à-vis competing actors –such as the United States and China–, while trying to influence the global debate from the "European way" with like-minded partners (EC, 2021a, 2021b, pp. 15–17).

Being a Team Europe Initiative, the EU-LAC Digital Alliance promotes a multi-stakeholder approach, including EU institutions and bodies, EU Member States, international financial institutions, the private sector, and civil society. It is embedded in the EU's Global Gateway investment strategy, whose main financial tool is the European Fund for Sustainable Development Plus established under the Neighborhood,

---

7 Discussion on launching a bi-regional digital alliance had taken place before, for instance, in the EU-LAC Informal Ministerial Meeting (Berlin, December 2020), the EU-LAC Leaders' Meeting (online, December 2021), and the third EU-CELAC Ministerial Meeting (Buenos Aires, October 2022).

Development, and International Cooperation Instrument (NDI-CI)-Global Europe,[8] to narrow the global investment gap in infrastructure using private and public resources (EC & HR, 2021; EC, 2021b, pp. 49–50). The Digital 4 Development Hub, which is another Team Europe Initiative but of global reach, ensures the coordination of the EU-LAC Digital Alliance among EU partners such as Luxembourg, France, Germany, Finland, Portugal and Spain.[9] The United Nations Economic Commission for LAC plays an equivalent role for partners in the LAC region (EC, 2023a).

The Digital Alliance puts special emphasis on privacy and personal data protection, drawing inspiration from the EU's General Data Protection Regulation,[10] among other areas of strong potential for cooperation (EC, 2021b, p. 16; EC & HR, 2021, p. 4). It also seeks to build on and advance existing EU-LAC digital cooperation, such as the Building the Europe Link to Latin America (BELLA) program, launched in the 2010s to provide ultra-high speed (inter)connectivity within Latin American and with the EU research and education communities, or the cooperation in the space sector under the EU Space Program (EC, 2021b, p. 16; 2023a, 2023b; EC & HR, 2021, p. 5; 2023, p. 9). For example, the BELLA II program had already started at the time of writing, and progress has been made in the creation of a regional center of the EU's Earth observation program –knows as Copernicus– in Panama (Directorate-General for International Partnerships –DG INTPA– 2022; Red Clara, 2023).

Different legally binding and non-binding instruments and tools, some of them already in place, complement the implementation of the EU-LAC Digital Alliance at the bilateral level. These include sectoral agreements for scientific and technological cooperation, as well as sectoral policy dialogues on matters pertaining to digital economy, cybersecurity, and information and communication technology (EC, 2023c). Another example are the chapters or titles on digital trade included in

---

8 OJ L 209/1, 14.6.2021. The Regional Multiannual Indicative Program for the Americas and the Caribbean (2021–2027), approved by the end of 2021, already envisaged the Digital Alliance as a priority area and a Team Europe Initiative to be supported within the Global Gateway (EC, 2021b, pp. 15–19, 49–50).

9 A tracker of the Team Europe Initiatives worldwide is available at: https://capacity4dev.europa.eu/resources/team-europe-tracker_en (Accessed: 8 December 2023).

10 OJ L 119/1, 4.5.2016; OJ L 127/2, 23.5.2018.

EU's international agreements as part of its trade policy, such as the agreement with Mexico that is in process to be updated at the time of writing.[11]

Arguably, the EP has (in)directly participated in shaping the EU-LAC Digital Alliance via legislative and non-legislative instruments, including its budgetary powers regarding the approval of the 2021–2027 Multiannual Financial Framework. The same holds true for the adoption of the EU's external financing instruments subject to the Ordinary Legislative Procedure, including the NDICI-Global Europe (Lilyanova, 2019). In the same vein, the EP can exert scrutiny over the actions of other EU institutions –mainly, the EC– but also over those of EU agencies and bodies (Corbett et al., 2016, pp. 365–366; Lilyanova, 2019, p. 10). The latter applies to the European Investment Bank, which is a major player in the implementation of the Global Gateway strategy (EC & HR, 2021, pp. 9–11). Following the entry into force of the Lisbon Treaty, the EP should also give consent to virtually any international agreement prior to its conclusion by the Council of the EU (Corbett et al., 2016, p. 299). This will be the case of the modernized agreement with Mexico, including their digital trade provisions, when the time comes.[12]

The next section examines how the EP has dealt with digital matters within EU-LAC relations through its PD, with special focus on the protection of privacy and personal data rights. PD understood here *stricto sensu*, that is as the EP's international relations *per se* (Jančić, 2017 p. 21) via the EuroLat.

4. The EuroLat in Action: Contributions to the EU-LAC Digital Agenda Regarding the Right to Privacy and Personal Data Protection

The EuroLat has adopted over 50 non-binding plenary resolutions and recommendations in the last decade.[13] Various of these documents

[11] On this modernization, see the policy papers co-edited by the European Institute of International Studies and the Konrad-Adenauer-Stiftung (Mexico), available at: https://www.ieeiweb.eu/publications (Accessed: 8 December 2023).

[12] On the role of the EP in the modernization of the EU-Mexico agreement, see Velasco-Pufleau (2023).

[13] Available at: https://www.europarl.europa.eu/eurolat/en/plenary/resolutions (Accessed: 8 December 2023).

refer to digital issues at some point in the text. A recent example is the resolution "Parliamentary scrutiny of trade negotiations and the evaluation and monitoring of free-trade agreements", adopted in the fifteenth Ordinary Plenary Session (Madrid, July 2023) (EuroLat, 2023g, para. 5).

Principally, two plenary resolutions address digital matters in a monographic way. The first one is named "Resolution on a Digital Agenda in the EU and the LAC countries" (EuroLat, 2016c) and was adopted based on a report by the Committee on Sustainable Development, Environment, Energy Policy, Research, Innovation and Technology in the ninth Ordinary Plenary Session (Montevideo, September 2016). The resolution dedicates a whole section to regulation and rights protection, containing specific recommendations to shape official relations from a parliamentary perspective. Among others, it urges both sides of the Atlantic to guarantee the right to privacy and the protection of personal data, including promoting the security of and confidence in Internet use, strengthening and harmonizing the EU and LAC countries' legal frameworks on consumer protection, and pursuing a common agenda at multilateral level that upholds human rights and fundamental values both online and offline (Ibidem paras. 27, 29–30). In the same vein, two joint declarations by the EuroLat's Co-presidents publicly supported the importance of implementing the new digital agenda ahead of and after the adoption of the resolution (EuroLat, 2016a, para. 17; 2016b, para. 21).

As suggested in the text of the 2016 resolution (EuroLat, 2016c, citations; see also EuroLat, 2015, pp. 4–6), its content heavily draws from policy documents adopted to boost the EU and LAC digital economies and societies, respectively. Among them is the Europe 2020 strategy that established a "Digital Agenda" for the EU itself to attain growth targets in the post-economic and financial crisis period of 2008, based on the strengthening of knowledge and innovation (EC, 2010, pp. 11–14). Still, the EuroLat's resolution remained innovative by introducing formal discussions on a digital agenda into the EU-LAC relations at the parliamentary level.

At the bi-regional level, the 2016 resolution was in general aligned with official developments on this field reflected at the time, for example, in the official documents of the first EU-CELAC Summit (Santiago de Chile, January 2013) (see Council of the EU, 2013a, para. 18; 2013b, section 1), which preceded the formal process for adoption of the EuroLat's resolution. This confirms previous findings that the "EuroLat's agenda is not entirely different from EU-LAC governmental dialogue" (Luciano, 2017, p. 326; see also Müller, 2019, p. 331), bringing digitalization as another area in which the Assembly and the Summits keep a relevant degree of thematic convergence. However, the EuroLat's resolution stressed the importance of privacy and data protection by introducing the whole above-mentioned section; an element that was missing in the official documents released by the first EU-CELAC Summit (Santiago de Chile, January 2013). Notably, this brings to light EuroLat's potential to also innovate at the level of EU-LAC official relations, by drawing attention to publicly unattended topics and asking authorities to incorporate them into further discussions. This parliamentary emphasis on privacy and data protection is far from surprising considering that the EP, in particular, has been a very strong supporter of these rights in EU law and policy over the years, translating its concerns to the EU's external relations (Mildebrath, 2023, pp. 7–8).

It is also relevant to address at this point the EuroLat's capacity to offer in-depth expert knowledge to the EU-LAC official relations, based on the professional background of its members in the field of digitalization. An example from the EU component is found in Carlos Zorrinho (Group of the Progressive Alliance of Socialists and Democrats, Portugal), who was co-rapporteur for the 2016 plenary resolution on the EU-LAC Digital Agenda.[14] Zorrinho is well known for his work in this field as a former government representative of Portugal and a member of the EP. Among other responsibilities, he led the development of the Portuguese Digital Agenda as National Coordinator for the Lisbon Strategy and the Technological Plan (2005–2009) and was State Secretary for Energy and Innovation (2009–2011) before becoming a member of the EP in 2014. As a member of this institution, he was rapporteur – appointed on 9 September 2014– of the Industry, Research and Energy Committee for the Program on interoperability solutions and common frameworks for European public administrations, businesses and citizens (ISA2 program) prior to the adoption of the EuroLat's resolution.[15] As with his previous work at the Portuguese government, this program was

[14] This example is used in light of the chapter's strong focus on the EP. Therefore, it does not intend to place less importance to the work and experience of the co-rapporteur from the LAC component but to provide evidence to support the claim made.

[15] Procedure 2014/0185(COD).

embedded in the challenges posed by the digital transition in the EU and its Member States. It could be highly expected that Zorrinho's professional experience – along with that of other EuroLat's members, including the co-rapporteur from the LAC component– would be incorporated into the resolution in question, which was meant to be transmitted to the Presidency of the EU-CELAC Summit, among other EU and LAC representatives (EuroLat, 2016c, para. 35).

The second monographic EuroLat's resolution dealing with digital matters is named "Digital platforms in the modern economy" (EuroLat, 2023h) and is based on a report by the Committee on Economic, Financial and Commercial Affairs. It was approved at committee level by the end of 2022 (EuroLat, 2022b, p. 2), when the EuroLat's Co-presidents also called for the green and digital transitions to be "equitable and inclusive" within the context of recovery from the COVID-19 pandemic (EuroLat, 2022a, para. 19). Eventually, the Assembly adopted the resolution in its fifteenth Ordinary Plenary Session (Madrid, July 2023), held after the third EU-CELAC Summit (Brussels, July 2023). Yet, its process of adoption can be traced back at least to May 2017, when it was agreed to launch a debate on the topic of digital platforms in the course of 2017 or 2018 (EuroLat, 2017, p. 4). This locates the origins of the resolution in a timeframe post-second EU-CELAC Summit (Brussels, June 2015) and prior to the third Summit (Brussels, July 2023). In fact, it is worth noting that the agreement took place during the meetings of the standing parliamentary committees that followed the ninth Ordinary Plenary Session in Montevideo (September 2016), when the 2016 resolution on the EU-LAC Digital Agenda was approved. This suggests that either the Committee on Sustainable Development, Environment, Energy Policy, Research, Innovation and Technology or the Committee on Economic, Financial and Commercial Affairs of the EuroLat have been working at varying capacities on these topics for years now with the goal of producing plenary resolutions.

As with the first EU-CELAC Summit (Santiago de Chile, January 2013), the second Summit (Brussels, June 2015) highlighted the importance of strengthening bi-regional cooperation regarding the digital economy and the reduction of the digital divide in its final declaration and action plan (EU-CELAC, 2015a, paras. 43 and 62; 2015b, section 1). However, it again omitted going deeper into the question of personal data and privacy protection. Conversely, the EuroLat's resolution on Digital platforms in the modern economy does expressly recognize the importance of these rights, calling to guarantee them as part of the rights and protection of consumers worldwide (EuroLat, 2023h, paras. 16–17). This importance of dealing with data and privacy protection in relation to consumers had already been underlined during the discussions of the working papers of the resolution in 2019 (EuroLat, 2019, pp. 2–4). Once more, this fact demonstrates the general thematic congruence between the official and parliamentary dimensions of EU-LAC relations. Yet, it also reaffirms the differences in the two approaches towards specific topics such as the protection of privacy and personal data. In this respect, it can be argued that the EuroLat has pursued a more "moral" approach by openly advocating for the upholding of these rights not only in the EU and the LAC region but globally.

Closely related to the previous point is the observation that the gap between both the parliamentary and the government approaches may evolve depending on the time frame of relations. In contrast to the second EU-CELAC Summit (Brussels, June 2015) and in consonance with the EuroLat's resolution on Digital platforms –adopted in committee in 2022 and at the plenary level in 2023–, the third Summit (Brussels, July 2023) did address the question of privacy in its official declaration. Specifically, it stresses "the importance of cooperating to promote a responsible human-centric, values-based and inclusive model of digital transformation that *protects privacy as a fundamental right*, increases digital connectivity and cybersecurity, aims at closing digital gaps, fosters the trustworthy development and use of Artificial Intelligence, and contributes to trust in the digital economy" (Council of the EU 2023, para. 29, emphasis added). This official position was also aligned with the EuroLat's Message to the third Summit (Brussels, July 2023) and the statement by the Co-presidency on its participation in this bi-regional event, both of which stress the need of intensifying governments' efforts to establish the conditions for a fair and inclusive digital transition (see EuroLat, 2023c, para. 36; 2023d).

Although further research is needed to confirm this observation, a possible explanation for changes in the governmental position over time could be related to the specific EU's strategies towards LAC surrounding each Summit. Indeed, the 2023 Joint Communication by the EC and the HR "A New Agenda for Relations between the European Union and Latin America and the Caribbean", adopted prior to the third Summit (Brussels, July 2023), envisages supporting partnerships for an inclusive and human-centric digital transformation as a priority. This includes through Global Gateway and international digital partnerships such as the EU-LAC Digital Alliance (EC & HR, 2023, pp. 6, 9–11).[16] Conversely, the 2009 Communication by the EC "The European Union and Latin America: Global Players in Partnership", applicable to previous summits, was silent regarding this topic (EC, 2009).

Besides the two monographic resolutions on digital matters cited above, the EuroLat's plenary and standing committees have organized various exchanges of views with government representatives and external experts dealing with these matters during the period of analysis of this chapter. These exchanges have given the opportunity to the EuroLat's members to gather relevant information and express their opinions directly to officials, constituting another way of providing parliamentary input to authorities for the development of bi-regional relations. An example is the session "Overcoming the consequences of the pandemic, supporting people and businesses: social and economic recovery" scheduled at the thirteenth Ordinary Plenary Session (Remote, November 2020) (EuroLat, 2020a, pp. 5–9). Following the plenary session, the Co-presidents also issued a joint declaration that heavily focused on the digital transition to address the consequences of the COVID-19 pandemic, urging authorities to cover the topic in further high-level bi-regional meetings (EuroLat, 2020b, para. 17). Without pretending to claim a direct causal link, since more research is needed to confirm this observation, it is true that the topic was indeed discussed in the EU-LAC Informal Ministerial Meeting held in Berlin some weeks later –on 14 December 2020–, with special attention to a future strategic digital partnership (European External Action Service –EEAS– 2020). A new communiqué by the EuroLat's Co-presidency was released after the meeting, endorsing the conclusions of governmental deliberations (EuroLat, 2020c). In this way, they provide democratic support and legitimacy to the initiative.

In the same context, it is worth noting the debate on "Reinforcing cooperation to protect people's digital rights" that took place in the

---

16 Before this 2023 Joint Communication, the 2019 Joint Communication "European Union, Latin America and the Caribbean: joining forces for a common future" referred to advancing the digital economy as an area to be strengthened by the EU with LAC partners, but without embracing the full idea of promoting an inclusive and human-centric digital partnership as a key area of relations. Still, this joint document highlighted the importance of enhancing convergence between EU and LAC on personal data protection (EC & HR, 2019, p. 5).

meeting of the Committee on Political Affairs, Security and Human Rights held within the framework of the fifteenth Ordinary Plenary Session (Madrid, July 2023), where two experts were invited (EuroLat, 2023e, pp. 3–4). Previously, members of the Committee on Sustainable Development, Environment, Energy Policy, Research, Innovation and Technology met with representatives of the EC's Directorate-General for Communications Networks, Content and Technology ahead of the adoption of the 2016 resolution on the Digital Agenda, to name but another example (EuroLat, 2015, pp. 4–6).

Moreover, there is evidence that the Euro-Latin American Women's Forum has discussed the challenges women face in the era of digitalization at least in 2018, incorporating a gender perspective into parliamentary debates (EuroLat, 2018b, p. 6). Similarly, the topic of digitalization has been part of EuroLat's meetings with civil society actors from the EU and LAC, allowing them to remain informed on bi-regional developments and express their viewpoints on digital matters outside of parliaments and governments. It is illustrative the "Exchange of views on the reduction of inequalities to promote sustainable development and the strengthening of democracies in Latin America and the Caribbean and in the European Union" held within the framework of the fifteenth Ordinary Plenary Session (Madrid, July 2023), when a DG INTPA representative spoke on the agreement reached at the third EU-CELAC Summit (Brussels, July 2023) on the Global Gateway investment agenda (EuroLat, 2023 f, p. 2).

Following the endorsement of the EU-LAC Digital Alliance by the relevant Heads of State and Government in the third Summit (Brussels, July 2023), Eurolat has paid special attention to the scrutiny of the inter-governmental initiative. The appearance of the DG INTPA representative before EuroLat's members and civil society actors has already been mentioned. More recently, this has been evident in the meetings of the EuroLat's standing parliamentary committees that took place in Santo Domingo (30 October-2 November 2023), where the topic "Digital and Educational Alliance" was one of the three axes framing discussions (EuroLat, 2023b). Remarkably, the so-called Santo Domingo Declaration adopted at the end of the meetings welcomes the EU-LAC Digital Alliance and the related investments foreseen under the EU's Global Gateway strategy (EuroLat, 2023i, para. 21), again providing publicly democratic support and legitimacy to the process.

6. Conclusions

The chapter has examined the opportunities for parliaments to engage in the promotion of the right to privacy and personal data protection in international relations, based on an in-depth understanding of the EuroLat's work and outcomes on digital matters over the last decade (2013–2023). Its insights seek to contribute to upholding these rights in the EU and its relations with the wider world, as well as advancing the literature on PD regarding digital matters.

Indeed, the chapter has shown that PD stands as an important tool for parliaments to address the challenges posed by the global digital transformation, which were exacerbated by the COVID-19 pandemic. In particular, PD granted them the possibility to cope with these challenges in a coordinated way beyond their own polities, through dialogue and collaboration.

The PD contributions to the protection of privacy and personal data in the EU and its partner countries were multiple. For example, it provided parliamentary input representing a wide range of political views, in-depth expert knowledge, and a gender perspective to EU-LAC inter-governmental exchanges; pushed government representatives from both regions to further address these rights in official discussions; advocated for a strong protection of these rights not only in the EU and LAC but globally, following a more moral approach to digitalization; contributed to disseminating information on official developments related to digital matters among parliamentarians and civil society actors from the EU and LAC countries, while allowing them to express concerns; and gave support and legitimation to inter-governmental initiatives, thereby adding a vital democratic layer to international digital governance.

The PD means used by EuroLat were in line with those traditionally available to parliaments in the field of foreign policy. Notably, they included plenary resolutions, committee reports, and exchanges of views with government representatives, external experts and civil society. Likewise, the Co-presidents and the Executive Bureau of the Assembly were particularly active in voicing EuroLat's concerns and recommendations via joint statements.

In sum, the findings demonstrate that parliaments can –and should play– a relevant role in upholding the right to privacy and personal data protection in the EU and beyond through PD. Therefore, the value of

PD in the digital age should not be underestimated but reinforced. All allies, including parliaments embracing liberal democratic values, should be welcomed in the endeavor of promoting a digital transformation that puts human rights at the center.

References

Bartels, L. (2014). The EU's human rights obligations in relation to policies with extraterritorial effects. *The European Journal of International Law, 25*(4), 1071–1091.

Corbett, R. (1998). *The European Parliament's role in EU closer integration.* New York: St. Martin's Press.

Corbett, R., Jacobs, F., & Neville, D. (2016). *The European Parliament* (9th ed.). London: John Harper Publishing.

Council of the EU. (2013a). *Santiago declaration,* Doc. No. 5747/13, Presse 31, Santiago de Chile.

Council of the EU. (2013b). *EU-CELAC action plan 2013–2015*, Doc. No. 5748/13, Presse 32, Santiago de Chile.

Council of the EU (2023). *Declaration of the EU-CELAC summit 2023.* Doc. No. 12000/23. Brussels: Council of the EU.

DG INTPA. (2022). *Global gateway in Latin America and the Caribbean: European Space Agency and the Commission join forces on earth observation.* https://international-partnerships.ec.europa.eu/news-and-events/news/global-gateway-latin-america-and-caribbean-european-space-agency-and-commission-join-forces-earth-2022-12-06_en. Accessed 8 December 2023.

EC. (2009). *Communication from the commission to the European Parliament and the Council—The European Union and Latin America: Global players in partnership*, Doc. No. COM(2009) 495 final, Brussels.

EC. (2010). *Communication from the commission – Europe 2020: A strategy for smart, sustainable and inclusive growth,* Doc. No. COM(2010) 2020 final, Brussels.

EC. (2021a). *Communication from the commission to the European Parliament, the Council, the European Economic and Social Committee and the Committee of the Regions—2030 Digital Compass: The European way for the digital decade,* Doc. No. COM(2021) 118 final, Brussels.

EC. (2021b). *The Americas and the Caribbean regional multiannual indicative programme 2021–2027*, https://international-partnerships.ec.europa.eu/system/files/2022-01/mip-2021-c2021-9356-americas-caribbean-annex_en.pdf. Accessed 8 December 2023.

EC. (2023a). *Global gateway: EU, Latin America and Caribbean partners launch in Colombia the EU-LAC digital alliance*, https://ec.europa.eu/commission/presscorner/detail/en/ip_23_1598. Accessed 8 December 2023.

EC. (2023b). *EU-Latin America and Caribbean: Joint declaration on a digital alliance.* https://ec.europa.eu/commission/presscorner/detail/en/statement_23_3892. Accessed 8 December 2023.

EC. (2023c). *Shaping Europe's digital future—Americas.* https://digital-strategy.ec.europa.eu/en/policies/americas. Accessed 8 December 2023.

EC & HR. (2019). *Joint Communication to the European Parliament and the Council—European Union, Latin America and the Caribbean: Joining forces for a common future*, Doc. No. JOIN(2019) 6 final, Brussels.

EC & HR. (2021). *Joint Communication to the European Parliament, the Council, the European Economic and Social Committee, the Committee of the Regions and the European Investment Bank – The Global Gateway*, Doc. No. JOIN(2021) 30 final, Brussels.

EC & HR. (2023). *Joint Communication to the European Parliament and the Council—A New Agenda for Relations between the EU and Latin America and the Caribbean*, Doc. No. JOIN(2023) 17 final, Brussels.

EEAS. (2020). *Joint Communiqué: EU27—Latin America and Caribbean Informal Ministerial Meeting*, https://www.eeas.europa.eu/eeas/joint-communiqu%C3%A9-eu27-latin-america-and-caribbean-informal-ministerial-meeting_en. Accessed 8 December 2023.

EP Conference of Presidents. (2015). *Decision on the Implementing provisions governing the work of delegations and missions outside the European Union*, Doc. No. PE422.560/CPG, Brussels.

EU-CELAC Summit. (2015a). *Brussels declaration*, Doc. No. 9839/15, Brussels.

EU-CELAC Summit. (2015b). *EU-CELAC action plan*. https://www.consilium.europa.eu/media/23757/eu-celac-action-plan.pdf. Accessed 8 December 2023.

EuroLat. (2015). *Minutes of the meeting of 4 June 2015 of the Committee on Sustainable Development, Environment, Energy Policy, Research, Innovation and Technology*, Doc. No. EUAL_PV(2015)0604, Brussels.

EuroLat. (2016a). *Lisbon declaration by the Co-Presidents of the Euro-Latin American Parliamentary Assembly (EuroLat) of 17 May 2016*, Doc. No. AP101.947v03-00, Lisbon.

EuroLat. (2016b). *Montevideo declaration by the Co-Presidents of the Euro-Latin American Parliamentary Assembly (EuroLat) of 21 September 2016*, Doc. No. AP102.137v03-00, Montevideo.

EuroLat. (2016c). *Resolution on a digital agenda in the EU and the LAC countries*, Doc. No. AP101.935v04-00, Montevideo.

EuroLat. (2017). *Minutes of the meeting of 23 May 2017 of the Committee on Economic, Financial and Commercial Affairs*, Doc. No. EUAL_PV(2017)0523_1, Florence.

EuroLat. (2018a). Rules of procedure, Doc. No. DV\1163861EN.docx, Brussels.

EuroLat. (2018b). *Minutes of the eleventh Ordinary Plenary Session, held on 19 September and 20 September 2018*, Doc. No. EUAL_PV(2019)0919_1, Vienna.

EuroLat. (2019). *Minutes of the meeting of 12 December 2019 of the Committee on Economic, Financial and Commercial Affairs*, Doc. No. AP102.926v01-00, Panama City.

EuroLat. (2020a). *Minutes of the Thirteenth Ordinary Plenary Session, held on 4 November 2020 and 5 November 2020*, Doc. No. EUAL_PV(2020)1104-05_01, Brussels.

EuroLat. (2020b). *Declaration of the Co-Presidents of the EuroLat on a comprehensive and bi-regional EU-LAC strategy to mitigate the impact of the COVID-19 pandemic, of 5 November 2020*, Doc. No. AP102.943v02-00, Brussels.

EuroLat. (2020c). *Communiqué by the Co-presidency of the EuroLat on its participation in the Informal Meeting of the Foreign Ministers of EU-27 and Latin America and the Caribbean (LAC)*, Doc. No. PE102.946v01-00, Berlin.

EuroLat. (2022a). *Declaration of the Co-Presidents of the EuroLat on 'Addressing the consequences of Russia's invasion of Ukraine with a view to strengthening peace and the multilateral order' of 30 November 2022*, Doc. No. AP103.221v02-00, Brussels.

EuroLat. (2022b). *Minutes of the meeting of 1 December 2022 of the Committee on Economic, Financial and Commercial Affairs*, Doc. No. EUAL_PV(2022)1201_1, Brussels.

EuroLat. (2023a). *Women's forum.* https://www.europarl.europa.eu/eurolat/en/working-groups-forums/womens-forum. Accessed 8 December 2023.

EuroLat. (2023b). *2023: Eurolat meetings.* https://www.europarl.europa.eu/eurolat/en/eurolat-meetings/eurolat-meetings.2023. Accessed 8 December 2023.

EuroLat. (2023c). *Message to the Third European Union (EU)—Community of Latin American and Caribbean States (CELAC) Summit*, Doc. No. PE100.144v03-00, Brussels.

EuroLat. (2023d). *Statement by the Co-presidency of the Eurolat on its participation in the third European Union (EU)—Community of Latin American And Caribbean States (CELAC) Summit*, Doc. No. AP103.451v01-00, Brussels.

EuroLat. (2023e). *Minutes of the meeting of 25 July 2023 of the Committee on Political Affairs, Security and Human Rights*, Doc. No. EUAL_PV(2023)0725, Madrid.

EuroLat. (2023f). *Minutes of the meeting of 26 July 2023 with civil society*, Doc. No. EUAL_PV(2023)0726_1, Madrid.

EuroLat. (2023g). *Resolution Parliamentary scrutiny of trade negotiations and the evaluation and monitoring of free-trade agreements*, Doc. No. AP103.229v03-00, Madrid.

EuroLat. (2023h). *Resolution Digital platforms in the modern economy*, Doc. No. AP103.130v05-00, Madrid.

EuroLat. (2023i). *Declaración Abordando viejos y nuevos retos para sentar las bases de una mayor prosperidad compartida en el futuro*, https://www.europarl.europa.eu/cmsdata/277991/Declaracion%20EuroLat_Santo%20Domingo.pdf. Accessed 8 December 2023.

Fernández Fernández, J. J. (2022). Pasado, Presente y Futuro de la Dimensión Parlamentaria de la Asociación Estratégica Birregional Unión Europea-América Latina: De las Conferencias Interparlamentarias a la Asamblea Parlamentaria Euro-Latinoamericana (EUROLAT). In C. F. Molina del Pozo (dir), *El Futuro de los Procesos de Integración en el marco de las Relaciones Unión Europea—América Latina* (pp. 373–391). Colex, A Coruña.

Jančić, D. (2017). World diplomacy of the European Parliament. In S. Stavridis & D. JanJančičić (eds.), *Parliamentary diplomacy in European and Global Governance* (pp. 19–40). Leiden: Brill.

Lamoso González, P. & Stavridis, S. (2022). El Parlamento Europeo como tribuna moral internacional: el caso de Venezuela. *Revista CIDOB d'Afers Internacionals*, 131, 185–210.

Lilyanova, V. (2019). *Financing EU external action in the new MFF, 2021–2027: Heading 6 'Neighbourhood and the World'*, Doc. No. PE644.173, EP Research Service, Brussels.

Luciano, B. T. (2017). Inter-parliamentary European Union–Latin American Caribbean (EU-LAC) relations and the increasing political convergence among Latin American regional parliaments (2006–15). *Parliaments, Estates and Representation*, 37(3), 318–334.

Manners, I. (2008). The normative ethics of the European Union. *International Affairs*, 84(1), 45–60.

Mildebrath, H. (2023). *Understanding EU data protection policy*, Doc. No. PE698.898, EP Research Service, Brussels.

Müller, G. G. (2019). The Euro-Latin America Parliamentary Assembly (EuroLat): constructing a common bi-regional agenda. In K. Raube et al. (eds.), *Parliamentary cooperation and diplomacy in EU external relations: An essential companion* (pp. 324–340). Cheltenham: Edward Elgar Publishing.

Red, Clara. (2023). *Bella II—Building the Europe Link to Latin America and the Caribbean.* https://www.redclara.net/index.php/es/proyectos/en-ejecucion/bella-ii. Accessed 8 December 2023.

Stavridis, S. (2017). Conclusions: Parliamentary diplomacy as a global phenomenon. In S. Stavridis & D. Jančić (eds.), *Parliamentary diplomacy in European and global governance* (pp. 368–387). Leiden: Brill.

Stavridis, S. (2019). La Diplomacia Parlamentaria: El papel de los parlamentos en el mundo. *Revista Española de Derecho Internacional*, 71(1), 187–206.

Stavridis, S. (2021). Parliamentary diplomacy: A review article. *International Journal of Parliamentary Studies*, 1, 227–269.

Stavridis, S. & Fernández Molina, I. (2013). El Parlamento Europeo y el conflicto de Libia (2011): ¿una tribuna moral eficiente?. *Revista CIDOB d'Afers Internationals*, 101, 153–176.

Velasco-Pufleau, M. (2017). The impact of parliamentary diplomacy, civil society and human rights advocacy on EU strategic partners: The case of Mexico. In S. Stavridis & D. Jančić (eds.), *Parliamentary diplomacy in European and global governance* (pp. 134–155). Leiden: Brill.

Velasco-Pufleau, M. (2023). Parliamentary diplomacy in practice: The role of the European Parliament delegations in the modernization of the global agreement between the European Union and Mexico. In P. Astroza Suarez et al. (eds.), *Over the Atlantic: diplomacy and paradiplomacy in EU and Latin America* (pp. 141–166). Lausanne: Peter Lang.

Wouters, J. & Ovádek, M. (2021). *The European Union and human rights: Analysis, cases, and materials.* Oxford: Oxford University Press.

Notes on Contributors

DR. PAULINA ASTROZA is a graduate lawyer from the University of Concepción, Chile. She holds a Ph.D. in Political and Social Sciences from the Université Catholique de Louvain, Belgium, with a thesis titled "Les activités internationales des régions chiliennes: embryon de diplomatie subétatique?" She completed a course in International Public Law at the Academy of International Law of the United Nations in The Hague, Netherlands. She also obtained a Diplôme d'Études Spécialisées (LLM) in Political Science and International Relations from the Université Catholique de Louvain, with a thesis titled "Les relations entre l'Union Européenne et le Chili: vers l'Association?" Additionally, she holds a Diplôme d'Études Complémentaires in International Relations and Comparative Politics from the same university. In addition, Dr. Astroza is the founder and director of the European Studies Center at the University of Concepción, Chile. She serves as a Professor of International Law, International Relations, and European Integration at the Faculty of Law and Social Sciences at the University of Concepción. Moreover, she is an Associate Professor in the Department of Public Law at the same institution and Jean Monnet Chair in 2017–2020; Jean Monnet Project 2018–2020 and Jean Monnet Project 2022–2025.

ALINE BELTRAME DE MOURA is Professor of Law School at the Federal University of Santa Catarina (UFSC), Brazil. Ph.D. in International Law from the Università degli Studi di Milano, Italy. Jean Monnet Chair on European Union Law. Co-Coordinator of the Jean Monnet Network Policy Debate "BRIDGE Watch", co-funded by the Erasmus + Programme. Chief-Editor of the Latin American Journal of European Studies and Director of the European Union Studies Chair at the European Institute of International Studies, Sweden.

GABRIEL HENRIQUE CERON TREVISOL is an Undergraduate Law Student at the Federal University of Santa Catarina. Researcher PIBIC/CNPq in the field of Law and Policies of the European Union. Member of the Latin-American Center of European Studies (UFSC). Member of the Research Group in Private International Law (NDIPr/UFSC).

CECILIA CELESTE DANESI is researcher at the Institute of European Studies and Human Right, Pontifical University of Salamanca, Spain. PhD. in AI & Law, University of Perugia (It) and Salamanca (Spain). Member of the platform "Women4EthicalAI" (UNESCO). Advisor in Digital area at ParlAméricas and in several governments. Visiting professor at the Universities of Oxford, Perugia, Uruguay Republic and Salamanca. Author of the book "The empire of algorithms" (among others) and publisher of several academic papers. Co-Director of the master's degree "Ethics Governance of AI" Pontifical University of Salamanca (Spain). Ambassador of the Women Economic Forum Argentina. Ted Talk Speaker.

MATTHIAS ERLANDSEN is PhD and MA in Communication Sciences from the Pontifical Catholic University of Chile, MA in International Studies from the University of Chile, and BA in Journalism and Social Communication from the Pontifical Catholic University of Chile. He is Chair of International Relations at FLACSO Chile, former Regional Lead for the Americas at the Office of the Vice-President, International at the University of Toronto (Canada), former Adjunct Lecturer at Universidad del Desarrollo (Chile).

ANTONIO NÚÑEZ Y GARCÍA-SAÚCO holds a doctorate in Philosophy and Humanities, a degree in Law and a degree in Political and Economic Sciences from the University of Madrid (UCM), completing postgraduate studies at Universities in Germany (Hamburg), France (Nancy and Strassbourg) and England (London University and Hollborn College). He was professor at the UCM, at the Spanish Diplomatic Academy and at the Center for Higher Studies of the National Defense of Spain (CESEDEN). He is a permanent collaborator at the Spanish Center for Strategic Studies (IEEE), where he has published several books. Until recently, he served as President of the European Institute of International Studies based in Salamanca and Stockholm (IEES-EIIS). As a member of the Spanish Diplomatic Service, he was Ambassador to, among others, the United Nations (Vienna), Australia, New Zealand and Sweden.

PASCAL GRISET is Professor of History in Sorbonne Université since 1998 and researcher at Sirice Institute. He is the coordinator of the H 2020 project "InsSCiDe" (Inventing a Shared Science Diplomacy for Europe). He is at the head of the Institute of Communication Sciences. Trained as a business historian, his research focusses on innovation, research policies and research organisations in France and Europe. Former Fulbright Visiting Fellow and auditor of the Institute for Industrial Strategy, he is currently president of the Scientific Committee for the history of INSERM (Institut national de la santé et de la recherche médicale), vice president of the Association for the History of Computing and Telecommunications, a member of the Committee for the economic and financial history of France, of the Scientific Committee for Historical Defense Research and the Scientific Council of the Committee for the History of La Poste.

MADELEINE O. HOSLI is a full Professor of Political Science (International Relations) at Leiden University in The Hague. Her main research interests are in international political economy, international organizations and European integration. She is author of The Euro: A Concise Introduction to European Monetary Integration (Lynne Rienner 2004), co-editor of Decision-Making in the European Union before and after the Lisbon Treaty (Routledge 2017), The Changing Global Order (Springer 2020), The Future of Multilateralism: Global Cooperation and International Organizations (Rowman and Littlefield 2021) and The Global, Regional and Local Politics of Institutional Responses to COVID-19: Implications for Women and Children (Palgrave MacMillan and Springer Nature 2022). She has published widely in peer-reviewed journals, including International Organization, International Studies Quarterly, the Journal of Common Market Studies, European Union Politics, the European Journal of Political Research and the Review of International Organizations.

MAGNUS LINDQVIST has worked as Senior Advisor and Project Coordinator for International Projects at KTH since 2012. He holds bachelor's and master's studies from Stockholm University and Södertörn University College, as well as from study exchange experiences from the University of the West Indies, in Jamaica. He is currently focused on international and externally funded research networks and projects, e.g. to develop, triple helix cooperation. He has primarily work experience from Europe and Latin America.

HANS LUNDIN has worked as Regional Advisor for Latin America at KTH Royal Institute of Technology since 2012. He holds a master's degree in political science at Umeå University and is currently also a PhD student at the HEOS group at KTH (Higher Educational Organisational Studies).

JONATHAN MICHAEL FELDMAN was born in New York City and has graduate degrees in regional economic development from MIT (master's in city planning) and Rutgers Universities (PhD). He was a Corliss Lamont Fellow in Economic Conversion and Disarmament at Columbia University in the United States and a Postdoctoral Fellow at Linköping University in Sweden. He presently works as an Associate Professor at Stockholm University and previously worked as Senior Fellow and Program Direction at the National Commission for Economic Conversion and Disarmament, based in Washington, D.C. headed by Professor Seymour Melman at Columbia University. Current research involves investigations of the Swedish-based Palestine International Business Forum, sustainability enablers in the fast fashion industry, and the mechanisms promoting and retarding green job innovation and development of rail-based passenger vehicles.

KAYLE VAN 'T KLOOSTER is a Dutch-Canadian scholar currently working as a junior lecturer at Leiden University. He teaches courses on a range of topics within the Bachelor Security Studies program. In addition to teaching, he has published works on EU-Chinese relations, Chinese-American relations and the European Indo-Pacific strategy. Previously he worked as a researcher for the Clingendael Institute. Kayle is a graduate of Leiden University, holding an MSc of the International Relations and Diplomacy program.

MARIO TORRES JARRÍN is Director of the Institute of European Studies and Human Rights at the Pontifical University of Salamanca (Spain). Previously, he was Researcher Associate and Adjunct Lecturer at Friedrich Alexander University Erlangen-Nuremberg (Germany), Lecturer and Research Associate in the Faculty of Humanities at Stockholm University and Director at the European Institute of International Studies (Sweden). He has been a Visiting Professor at Copenhagen Business School (Denmark), University of Bergen and University of Oslo (Norway), University Institute of Lisbon (Portugal), National University of Political Studies and Public Administration (Romania), University of Salamanca (Spain), University of Economics and Anglo-American University (Czech Republic) and Diplomatic Academics in Europe, Asia and Americas. He holds a PhD in History, a Master in European Union Studies, and a BA in Business Studies from the University of Salamanca (Spain).

MÓNICA VELASCO-PUFLEAU is Director of the European Union-Latin American and the Caribbean Relations Chair at the European Institute of International Studies since March 2021. She holds a PhD in Law-International Studies from the University of Barcelona (Spain) and has been Postdoctoral Research Fellow at the University of Luxembourg (Luxembourg). Her research interests include the role of the European Parliament in upholding fundamental rights in the European Union and its relations with third countries, parliamentary engagement on human rights issues, parliamentary diplomacy, privacy and personal data protection, and European Union external relations towards Latin America and the Caribbean.

www.ingramcontent.com/pod-product-compliance
Ingram Content Group UK Ltd.
Pitfield, Milton Keynes, MK11 3LW, UK
UKHW020836290926
15470UKWH00004B/1377

9 783631 926970